ALGEBRA FOR BEGINNERS

Feels like a friend explaining it to you

I don’t get this algebra stuff. I barely know anything.

Ugh... where do I even start?

I’ve tried. I really have.

Reread my textbook several times.

Took my own notes. Color-coded everything.

Watched YouTube videos. Even got a tutor.

Nothing worked.

It just doesn't make sense to me.

Like everyone else just... gets it.

And I'm still stuck on fractions.

I don't know... maybe I'm just bad at math.

Nobody’s bad at maths. It’s just taught badly.

And what you’re feeling?

Totally normal.

But don’t worry.

We’ll figure it out together.

You’ll get it. Promise.

You really think so?

Yeah. We’ll get you passing.

All you need is basic math.

Can you add, subtract, multiply, and divide?

Yes?

Perfect.

We’ll start from there and build up.

Here's what we'll do.

We go back to the basics.

The stuff that wasn't explained properly the first time.

From there, we rebuild.

Step by step.

One idea at a time.

In plain English.

Cuz chances are, somewhere along the way.

Something got missed.

Something didn't make sense.

And that missing thing?

That's what made everything else after it so confusing.

But we'll fix that.

Everything will be explained clearly.

No skipping steps.

You'll finally see how it all fits together.

That'd be... amazing.

I just... I wanna finally get this.

I know I've got the potential to be good at math.

You do.

And you will.

Let's do this.

Stuff You're About to Get Real Good At

Linear Equations

Simultaneous Equations

The Basics

Divisibility Rules: The Math Shortcuts No One Tells You About

Do you know what divisibility rules are?

Yes?

No?

Not sure?

Okay, so they're a shortcut to check if one number goes into another.

Without doing any division.

If you know'em you'll be able to simplify fractions way faster.

Or find common factors of big numbers.

You know how useful that is?

Cuz you do this stuff in pretty much every math class.

At any level.

So, it's definitely something you want to know.

Or at least be familiar with.

For some reason, they don't teach this in schools.

There's really only 9 main ones you need to know.

Here they are.

9 Divisibility Rules: Don't Skip This - It's Actually Pretty Cool

Before we get into the 9 rules, just a couple of things.

One.

These rules don't actually do the division for you.

They just tell you if one number divides evenly into another.

Two.

There *are* more rules for bigger numbers.

But these 9 are the ones you'll actually use.

For bigger numbers, you'll probably be allowed to use a calculator anyway.

Three.

The word "digit".

It just means a single number from 0 to 9.

Alright.

Here they are.

A Number is Divisible By...

2	If the last digit is even or 0
3	If the sum of the number's digits is divisible by 3
4	If the number formed by the last 2 digits is divisible of 4
5	If the last digit is 0 or 5
6	If the number is even AND divisible by 3
7	If double the last digit subtracted from the number formed by the rest of the digits is 0 or divisible by 7
8	If the number formed by the last 3 digits is a divisible of 8
9	If the sum of the digits is divisible by 9
10	If the last digit is 0

Worked Example 1

Check the divisibility of 66 by the following numbers:

a **2** b **3** c **4**

d **5** e **6**

Explanation		**Solution**
a	Look at the last digit.	The last digit is 6.
	Is it even?	6 is even.
	Answer.	∴ 66 is divisible by 2.
b	Find the sum of the digits.	$6 + 6 = 12$
	Is the sum divisible by 3?	12 is divisible by 3.
	Answer.	∴ 66 is divisible by 3.
c	Look at the last two digits of the number.	The last two digits are 66.
	Is it divisible by 4?	66 is NOT divisible by 4.
	Answer.	∴ 66 is NOT divisible by 4.
d	Look at the last digit.	The last digit is 6.
	Is it a 0 or 5?	It's not 0 or 5.
	Answer.	∴ 66 is NOT divisible by 5.
e	Is the last digit even?	6 is even.
	Is the number divisible by 3?	66 is also divisible by 3.
	Answer.	∴ 66 is divisible by 6.

Worked Example 2

Check the divisibility of 90 by the following numbers:

a **4**

b **5**

c **6**

d **7**

e **10**

Explanation		Solution
a	Look at the last two digits of the number.	The last two digits are 90.
	Is it divisible by 4?	90 is NOT divisible by 4.
	Answer.	∴ 90 is NOT divisible by 4.
b	Look at the last digit.	The last digit is 0.
	Is it a 0 or 5?	Yes.
	Answer.	∴ 90 is divisible by 5.
c	Is the last digit 0 or even?	Last digit is 0.
	Is the number divisible by 3?	90 is also divisible by 3.
	Answer.	∴ 90 is divisible by 6.
d	Double the last digit and subtract it from the number formed by the rest of the digits.	$9 - (2 \times 0) = 9$
	Is this number 0 or divisible by 7?	9 is NOT divisible by 7.
	Answer.	∴ 90 is NOT divisible by 7.
e	Is the last digit 0.	Yes
	Answer.	∴ 90 is divisible by 10.

Worked Example 3

Check the divisibility of 336 by the following numbers:

a **7** b **8** c **9**

Explanation	Solution
a Double the last digit and subtract it from the number formed by the rest of the digits.	$33 - (2 \times 6) = 21$
Is this number 0 or divisible by 7?	21 is divisible by 7.
Answer.	$\therefore$ 336 is divisible by 7.
b Divide the last 3 digits by 8 as normal.	$336 \div 8 = 42$
Does it divide evenly?	Yes.
Answer.	$\therefore$ 336 is divisible by 8.
c Find the sum of the digits.	$3 + 3 + 6 = 12$
Is the sum divisible by 9?	12 is NOT divisible by 9.
Answer.	$\therefore$ 336 is NOT divisible by 9.

Even numbers divisible by 4 aren't always divisible by 8.
For 3-digit numbers, use long division to check divisibility by 8.

Your Turn To Try

1 **Which of the following is divisible by 3?**

A 714　　B 827

C 263　　D 584

2 **Which of the given numbers is divisible by 2, 5 and 10?**

A 55　　B 432

C 120　　D 855

3 **For which of the following numbers is the rule for divisibility by 6 satisfied?**

A 370　　B 284

C 141　　D 528

4 **Which digit should replace the blank in $4 - 8$ to make the number divisible by 9?**

A 1　　B 6

C 3　　D 5

5 **Which digit can replace the blank in -36 to make the number divisible by 4?**

A 2　　B 5

C 7　　D *All of the above*

Question	1	2	3	4	5
Answer	A	C	D	B	D

Time To Practice

1 Is 24 divisible by 2?

2 Is 42 divisible by 3?

3 Is 45 divisible by 5?

4 Is 68 divisible by 6?

5 Is 72 divisible by 8?

6 Is 84 divisible by 6?

7 Is 84 divisible by 7?

8 Is 94 divisible by 7?

9 Is 98 divisible by 4?

10 Is 105 divisible by 9?

11 Is 114 divisible by 7?

12 Is 144 divisible by 4?

13 Which numbers from 2 to 10 are the following numbers divisible by?

a 120 b 156 c 198

d 240 e 357 f 378

g 434 h 480 i 496

j 504 k 739 l 840

Challenge Yourself

1 A three-digit number 3__8 is divisible by 3 but not by 9. What could be the missing digit?

2 A three-digit number 72__ is divisible by 4 and 7. What is the missing digit?

3 A three-digit number __50 is divisible by 2, 5, and 7. What is the smallest possible value of the missing digit?

4 A three-digit number 6__3 is divisible by 3, 7 and 9. What is the missing digit?

5 A three-digit number __04 is divisible by 6, 7 and 8. What is the missing digit?

Simplifying Fractions: Yeah, You Gotta Do It. It's Expected

Alright, simplifying fractions.

A fraction has two numbers.

Top number — that's called the *numerator*.

Bottom number — that's the *denominator*.

Simplifying means making both numbers smaller.

Without changing what the fraction equals.

How do we do that?

This is where divisibility rules come in.

Look at the fraction and ask yourself,

"What's the biggest number that goes into both the top and bottom?"

By the way, that number is called the Greatest Common Factor.

Or GCF for short.

Once you've got the GCF, you can simplify in two ways.

The quicker way.

Divide the top and bottom numbers by the GCF.

What's left is the simplified fraction.

The other way.

Rewrite the top and bottom numbers as two numbers being multiplied.

With one of those numbers being the GCF.

Then, cancel out the GCF from both the top and the bottom.

What's left is the simplified fraction.

By the way.

Cancel out means removing the same number from the top and bottom of the fraction.

It's also known as *crossing out*.

You just cross out the same numbers.

As if they weren't there.

OK, if we cancel out numbers, doesn't that change the fraction?

Nope, it doesn't.

Two fractions can look different but still be equal.

That's the point of simplifying.

We're rewriting the fraction using smaller numbers.

While keeping its value the same.

Does it matter which way you simplify?

Not really.

Both ways work.

It's more about which way will be easier.

If it's a single fraction with just numbers.

Usually faster to divide the top and bottom by the GCF.

But if you're multiplying fractions.

Or later on when you see fractions with letters or expressions,

the cancel-out method tends to be easier.

So, you wanna be good with both ways.

Anyway.

Not all fractions can be simplified.

If the top and bottom numbers don't have any numbers in common.

Except for 1.

Then the fraction's already as simple as it gets.

You don't have to do anything to it.

Just leave it as it is.

Now if the fraction has big numbers.

The GCF isn't always obvious.

So sometimes, you might end up simplifying more than once.

You'll simplify and notice the top and bottom still have a GCF.

Totally normal.

Happens all the time.

Just simplify again.

Try to get into the habit of always double-checking your answer.

Make sure the final fraction can't be simplified any further.

Because if it can, and you don't.

Your answer isn't complete.

And you could lose marks.

Alright, here are the steps to simplify a fraction.

1) Rewrite the top and bottom as two numbers multiplied together. One being the GCF
2) Cancel out the GCF from top and bottom
3) What's left is the simplified fraction

Worked Example 1

Simplify these fractions:

a $\frac{4}{12}$ b $\frac{5}{25}$ c $\frac{3}{27}$

Explanation		Solution
a	Write the fraction.	$\frac{4}{12}$
	Write the top and bottom numbers as multiples of the GCF.	$= \frac{4 \times 1}{4 \times 3}$
	Cancel out the common factor.	$= \frac{1}{3}$
b	Write the fraction.	$\frac{5}{25}$
	Write the top and bottom numbers as multiples of the GCF.	$= \frac{5 \times 1}{5 \times 5}$
	Cancel out the common factor.	$= \frac{1}{5}$
c	Write the fraction.	$\frac{3}{27}$
	Write the top and bottom numbers as multiples of the GCF.	$= \frac{3 \times 1}{3 \times 9}$
	Cancel out the common factor.	$= \frac{1}{9}$

Worked Example 2

Reduce the following fractions to its simplest form.

a $\frac{40}{90}$ b $\frac{30}{42}$ c $\frac{28}{50}$

Explanation	Solution
a Write the fraction.	$\frac{40}{90}$
Write the top and bottom numbers as multiples of the GCF.	$= \frac{10 \times 4}{10 \times 9}$
Cancel out the common factor.	$= \frac{4}{9}$
b Write the fraction.	$\frac{30}{42}$
Write the top and bottom numbers as multiples of the GCF.	$= \frac{6 \times 5}{6 \times 7}$
Cancel out the common factor.	$= \frac{5}{7}$
c Write the fraction.	$\frac{28}{50}$
Write the top and bottom numbers as multiples of the GCF.	$= \frac{2 \times 14}{2 \times 25}$
Cancel out the common factor.	$= \frac{14}{25}$

Your Turn To Try

1 **Which of these fractions is equal to $\frac{1}{2}$?**

A $\frac{10}{22}$

B $\frac{9}{18}$

C $\frac{11}{24}$

D $\frac{3}{9}$

2 **Which of these fractions is the same as $\frac{4}{5}$?**

A $\frac{12}{15}$

B $\frac{45}{55}$

C $\frac{20}{30}$

D $\frac{24}{25}$

3 **When simplified $\frac{6}{54}$ is?**

A $\frac{6}{7}$

B $\frac{4}{9}$

C $\frac{1}{9}$

D $\frac{5}{9}$

4 **What is $\frac{42}{48}$ in its simplest form?**

A $\frac{6}{7}$

B $\frac{7}{8}$

C $\frac{3}{8}$

D $\frac{3}{7}$

5 **Which one of these is NOT equal to $\frac{3}{4}$?**

A $\frac{12}{16}$

B $\frac{21}{28}$

C $\frac{33}{44}$

D $\frac{20}{24}$

Question	1	2	3	4	5
Answer	B	A	C	B	D

Time To Practice

1 Write the following fractions in its simplest form.

a $\frac{4}{36}$ b $\frac{5}{35}$ c $\frac{6}{18}$

2 Simplify the following fractions.

a $\frac{9}{63}$ b $\frac{8}{64}$ c $\frac{7}{77}$

3 Reduce the following fractions to its simplest form.

a $\frac{14}{21}$ b $\frac{16}{40}$ c $\frac{18}{81}$

4 Find the simplest form of:

a $\frac{21}{35}$ b $\frac{22}{55}$ c $\frac{24}{36}$

5 Reduce each fraction to its simplest form.

a $\frac{28}{63}$ b $\frac{30}{35}$ c $\frac{49}{56}$

6 Find the simplest form of the following.

a $\frac{50}{80}$ b $\frac{55}{99}$ c $\frac{60}{72}$

7 Simplify each fraction to its lowest terms.

a $\frac{72}{99}$ b $\frac{84}{108}$ c $\frac{132}{144}$

Challenge Yourself

1 Write the following fractions in its simplest form.

a $\frac{80}{120}$ b $\frac{54}{72}$ c $\frac{88}{110}$

2 Simplify the following.

a $\frac{120}{144}$ b $\frac{125}{175}$ c $\frac{300}{330}$

Adding and Subtracting Fractions: Don't Run Away Just Yet

If you want to be good at math, you have to be good at fractions.

No way around it.

Because fractions are everywhere in math.

In any topic.

At any level.

You're always going to see fractions.

And if you don't know how to work with them.

You will struggle.

It's not just simplifying you need to be good at.

You need to be good at adding, subtracting, multiplying, and dividing them too.

So, let's look at them one by one.

Starting with adding and subtracting.

Here are the rules.

Fraction Rules

Adding with Same Denominator	$\frac{a}{b}+\frac{c}{b}=\frac{a+c}{b}$
Subtracting with Same Denominator	$\frac{a}{b}-\frac{c}{b}=\frac{a-c}{b}$
Adding with Different Denominators	$\frac{a}{b}+\frac{c}{d}=\frac{ad+bc}{bd}$
Subtracting with Different Denominators	$\frac{a}{b}-\frac{c}{d}=\frac{ad-bc}{bd}$

Those letters a, b, c, and d.

They each represent a number.

To add or subtract fractions, the denominators have to be the same.

If they are, easy.

Add or subtract the numerators.

Keep the denominator the same.

If the denominators are different, you can't just add or subtract the numerators.

That only works when the denominators are the same.

So, what you do is this.

Either make the denominators the same – we'll go over how to do that next.

Or you cross-multiply.

Which is what the rules show.

Here's how to cross-multiply.

1) Multiply the denominators — that becomes the answer's denominator
2) To cross-multiply,
    i. Multiply the top of the first fraction by the bottom of the second
    ii. Multiply the top of the second fraction by the bottom of the first
3) Add or subtract those two results — that becomes the answer's numerator
4) Simplify if needed

When working with fractions, whole numbers often come up too.

Here's how to rewrite them as fractions.

Any number written over one equals that number.

$$a = \frac{a}{1}$$

Any number written over itself equals one.

$$1 = \frac{a}{a}$$

Worked Example 1

Calculate the following:

a $\frac{2}{7}+\frac{3}{7}$ b $\frac{5}{9}-\frac{2}{9}$ c $\frac{4}{11}+\frac{6}{11}$

Explanation	Solution
a Write the fractions.	$\frac{2}{7}+\frac{3}{7}$
The denominators are the same so add the numerators.	$=\frac{2+3}{7}$
(Fraction can't be simplified any further)	$=\frac{5}{7}$
b Write the fractions.	$\frac{5}{9}-\frac{2}{9}$
The denominators are the same so subtract the numerators.	$=\frac{5-2}{9}$
Simplify.	$=\frac{3}{9}$
	$=\frac{1}{3}$
c Write the fractions.	$\frac{4}{11}+\frac{6}{11}$
The denominators are the same so add the numerators.	$=\frac{4+6}{11}$
(Fraction can't be simplified any further)	$=\frac{10}{11}$

Worked Example 2

Calculate the following:

a $\frac{2}{3}+\frac{1}{5}$

b $\frac{5}{8}-\frac{1}{3}$

Explanation	Solution
a Write the fractions.	$\frac{2}{3}+\frac{1}{5}$
Multiply the denominators and write as a single fraction.	$\frac{2}{3}+\frac{1}{5}=\frac{}{15}$
Cross multiply the fractions: multiply each numerator by the denominator of the other fraction.	$=\frac{(2\times 5)+(1\times 3)}{15}$
Add the numerators.	$=\frac{10+3}{15}$
	$=\frac{13}{15}$
b Write the fractions.	$\frac{5}{8}-\frac{1}{3}$
Multiply the denominators and write as a single fraction.	$\frac{5}{8}-\frac{1}{3}=\frac{}{24}$
Cross multiply the fractions: multiply each numerator by the denominator of the other fraction.	$=\frac{(5\times 3)-(1\times 8)}{24}$
Subtract the numerators.	$=\frac{15-8}{24}$
	$=\frac{7}{24}$

Worked Example 3

Work out: $\dfrac{2}{9}+\dfrac{3}{8}$

Explanation	Solution
Write the fractions.	$\dfrac{2}{9}+\dfrac{3}{8}$
Multiply the denominators and write as a single fraction.	$\dfrac{2}{9}+\dfrac{3}{8}=\dfrac{}{72}$
Cross multiply the fractions: multiply each numerator by the denominator of the other fraction.	$=\dfrac{(2\times 8)+(3\times 9)}{72}$
Add the numerators.	$=\dfrac{16+27}{72}$
(Fraction can't be simplified any further)	$=\dfrac{43}{72}$

Adding or subtracting numerators when the denominators are different.

Worked Example 4

Calculate the following:

a $\frac{2}{5}+3$ b $7-\frac{1}{4}$

Explanation	Solution
a Write the fractions.	$\frac{2}{5}+3$
Write the whole number as a fraction with denominator 1.	$\frac{2}{5}+\frac{3}{1}$
Multiply the denominators and write as a single fraction.	$\frac{2}{5}+\frac{3}{1}=\frac{}{5}$
Cross multiply the fractions.	$=\frac{(2\times 1)+(3\times 5)}{5}$
Add the numerators.	$=\frac{2+15}{5}=\frac{17}{5}$
b Write the fractions.	$7-\frac{1}{4}$
Write the whole number as a fraction with denominator 1.	$\frac{7}{1}-\frac{1}{4}$
Multiply the denominators and write as a single fraction.	$\frac{7}{1}-\frac{1}{4}=\frac{}{4}$
Cross multiply the fractions.	$=\frac{(7\times 4)-(1\times 1)}{4}$
Subtract the numerators.	$=\frac{28-1}{4}=\frac{27}{4}$

Worked Example 5

Evaluate: $\mathbf{-5-\frac{3}{4}}$

Explanation	**Solution**
Write the fractions.	$-5-\frac{3}{4}$
Write the whole number as a fraction with denominator 1.	$-\frac{5}{1}-\frac{3}{4}$
Multiply the denominators and write as a single fraction.	$-\frac{5}{1}-\frac{3}{4}=\frac{\quad}{4}$
Cross multiply the fractions.	$=\frac{(-5\times 4)-(3\times 1)}{4}$
Subtract the numerators.	$=\frac{-20-3}{4}$
	$=-\frac{23}{4}$

If there's a negative in front of a fraction, it's the same as having a negative numerator or a negative denominator. But not both.

Worked Example 6

Work out: $\mathbf{1 - \left(\frac{1}{3} + \frac{1}{5}\right)}$

Explanation	Solution
Start with the expression.	$1 - \left(\frac{1}{3} + \frac{1}{5}\right)$
Add fractions inside brackets by cross multiplying.	$= 1 - \left(\frac{5+3}{15}\right)$
Simply the fraction inside the brackets.	$= 1 - \frac{8}{15}$
Write the number 1 as a fraction with the numerator and denominator as 15.	$= \frac{15}{15} - \frac{8}{15}$
Subtract the numerators.	$= \frac{15-8}{15}$
	$= \frac{7}{15}$

If you see fractions inside brackets, simplify them first. Remember BOMDAS?

Worked Example 7

Work out: $2 - \left(\frac{1}{7} - \frac{2}{5}\right)$

Explanation	Solution
Start with the expression.	$2 - \left(\frac{1}{7} - \frac{2}{5}\right)$
Subtract fractions inside brackets by cross multiplying.	$= 2 - \left(\frac{5 - 14}{35}\right)$
Simply the fraction inside the brackets. (Be careful with the negative inside brackets)	$= 2 - \left(-\frac{9}{35}\right)$
	$= 2 + \frac{9}{35}$
Write the whole number as a fraction with denominator 1.	$= \frac{2}{1} + \frac{9}{35}$
Cross multiply the fractions.	$= \frac{70 + 9}{35}$
Add the numerators.	$= \frac{79}{35}$

After adding or subtracting fractions, forgetting to simplify the answer. ⚠

Cross-Multiplying? Wait — Try This First

We cross-multiply when the denominators are different.

But sometimes, we don't need to do that.

It can make things a little bit harder.

Especially with big denominators.

So, before you cross-multiply, pause for sec.

Look at the denominators.

See if the smaller one goes into the bigger one.

If it does — sweet.

No need to cross-multiply.

There's an easier way to make the denominators the same.

Here's what to do.

Find what number to multiply the smaller denominator by so it equals the bigger one.

Multiply the top and bottom of the fraction with the smaller denominator by this number.

That'll make the denominators the same.

Once they're the same.

Add or subtract the numerators, and keep the denominator as is.

Then, simplify if possible.

Here are the steps in short.

1) Find what number to multiply the smaller denominator by to get the bigger one
2) Multiply the top and bottom of the fraction, with the smaller denominator, by this number
3) Add or subtract the numerators
4) Simplify if possible.

Worked Example 8

Calculate the following:

a $\frac{1}{3}+\frac{6}{21}$ b $\frac{4}{5}-\frac{1}{20}$

Explanation	Solution
a Write the fractions.	$\frac{1}{3}+\frac{6}{21}$
Multiply top and bottom of first fraction to match second denominator.	$=\frac{1\times 7}{3\times 7}+\frac{6}{21}$
Both fractions now have same denominator.	$=\frac{7}{21}+\frac{6}{21}$
Add numerators.	$=\frac{13}{21}$
b Write the fractions.	$\frac{4}{5}-\frac{1}{20}$
Multiply top and bottom of first fraction to match second denominator.	$=\frac{4\times 4}{5\times 4}-\frac{1}{20}$
Both fractions now have same denominator.	$=\frac{16}{20}-\frac{1}{20}$
Subtract numerators.	$=\frac{15}{20}$
Simplify.	$=\frac{3\times 5}{4\times 5}$
	$=\frac{3}{4}$

Worked Example 9

Calculate the following:

a $\frac{7}{15} - \frac{1}{5}$ c $\frac{21}{50} - \frac{3}{100}$

Explanation	**Solution**
a Write the fractions.	$\frac{7}{15} - \frac{1}{5}$
Multiply top and bottom of second fraction to match first denominator.	$= \frac{7}{15} - \frac{1 \times 3}{5 \times 3}$
Both fractions now have same denominator.	$= -\frac{3}{15} + \frac{7}{15}$
Subtract numerators.	$= \frac{4}{15}$
b Write the fractions.	$\frac{21}{50} - \frac{3}{100}$
Multiply top and bottom of first fraction to match second denominator.	$= \frac{21 \times 2}{50 \times 2} - \frac{3}{100}$
Both fractions now have same denominator.	$= \frac{42}{100} - \frac{3}{100}$
Subtract numerators.	$= \frac{39}{100}$

To find the number to multiply the fraction with the smaller denominator, divide the bigger denominator by the smaller one.

Your Turn To Try

1 What is $\frac{4}{5} - \frac{2}{3}$?

A $-\frac{2}{15}$

B $\frac{7}{15}$

C $-\frac{1}{3}$

D $\frac{2}{15}$

2 What is $\frac{11}{12} - \frac{5}{6}$?

A $\frac{1}{12}$

B 1

C $\frac{1}{6}$

D $\frac{1}{4}$

3 When simplified $\frac{7}{10} - 1$ is?

A $\frac{7}{5}$

B $-\frac{3}{10}$

C $-\frac{7}{10}$

D $\frac{3}{10}$

4 What is the simplified form of $1 + \left(\frac{3}{5} - \frac{2}{7}\right)$?

A $\frac{8}{7}$

B $\frac{46}{35}$

C $\frac{4}{35}$

D $\frac{66}{35}$

5 After simplifying $\frac{2}{3} - \left(\frac{5}{6} - \frac{1}{4}\right)$ which is the correct answer?

A $\frac{7}{12}$

B $-\frac{7}{12}$

C $\frac{1}{12}$

D $\frac{1}{4}$

Question	1	2	3	4	5
Answer	D	A	B	B	C

Time To Practice

1 Simplify the following.

a $\frac{1}{4}+\frac{2}{4}$

b $\frac{3}{8}-\frac{1}{8}$

c $\frac{1}{10}+\frac{9}{10}$

2 Work out the following as a simplified fraction.

a $\frac{5}{12}-\frac{1}{4}$

b $-\frac{7}{10}+\frac{4}{5}$

c $\frac{2}{3}-\frac{5}{6}$

3 Calculate.

a $\frac{1}{8}+\frac{3}{4}$

b $\frac{4}{7}+\frac{3}{10}$

c $\frac{9}{11}-\frac{3}{8}$

4 Simplify.

a $2-\frac{1}{4}$

b $-1+\frac{1}{6}$

c $3+\frac{7}{10}$

5 Evaluate each fraction.

a $1+\left(\frac{7}{24}-\frac{5}{12}\right)$

b $-1+\left(2-\frac{1}{2}\right)$

c $3-\left(\frac{9}{10}+\frac{2}{5}\right)$

6 Write each answer in lowest terms.

a $\left(\frac{1}{4}+\frac{2}{5}\right)-\frac{1}{10}$

b $-\frac{4}{9}+\left(\frac{1}{3}-\frac{1}{12}\right)$

c $\left(-\frac{1}{6}+\frac{1}{4}\right)+\frac{5}{8}$

7 Simplify each fraction to its lowest terms.

a $\left(-\frac{7}{10}-\frac{2}{3}\right)-\frac{1}{5}$

b $\left(-\frac{7}{10}-\frac{2}{3}\right)-\frac{1}{5}$

c $\left(-\frac{2}{3}-\frac{4}{5}\right)+\frac{2}{3}$

Challenge Yourself

1 Write the following fractions in its simplest form.

a $\left(\frac{8}{13}+\frac{1}{2}\right)-\frac{3}{26}$

b $-\frac{11}{20}+\left(2-\frac{1}{2}\right)$

2 Simplify the following.

a $\frac{1}{3}-\left(\frac{5}{6}-\frac{7}{18}\right)$

b $-\frac{2}{5}-\left(\frac{1}{10}-\frac{1}{15}\right)$

3 Calculate the result of the following.

a $\frac{2}{18}-\left(\frac{2}{9}-\frac{4}{27}\right)$

b $\left(\frac{5}{8}-\frac{1}{10}\right)-\frac{3}{4}$

Multiplying and Dividing Fractions: KFC (No, Not The Chicken Place)

Now let's look at multiplying and dividing fractions.

They're a little bit easier to work with.

Here are the rules.

Fraction Rules

Multiplying

$$\frac{a}{b} \times \frac{c}{d} = \frac{ac}{bd}$$

or

$$a \times \frac{c}{d} = \frac{ac}{d}$$

Dividing

$$\frac{a}{b} \div \frac{c}{d} = \frac{a}{b} \times \frac{d}{c} = \frac{ad}{bc}$$

or

$$a \div \frac{c}{d} = a \times \frac{d}{c} = \frac{ad}{c}$$

or

$$\frac{a}{b} \div c = \frac{a}{b} \times \frac{1}{c} = \frac{a}{bc}$$

When multiplying fractions, it doesn't matter if the denominators are the same or not.

Just multiply straight across.

Numerator by numerator.

Denominator by denominator.

Then simplify if needed.

If you have a whole number multiplied by a fraction.

Just multiply the whole number by the numerator.

Keep the denominator as it is.

Dividing fractions is just as easy.

It's basically the same as multiplying.

Except you do this flipping thing first.

You flip the second fraction.

Change the division sign to multiplication.

Then multiply straight across.

To remember this, think KFC.

No. Not the chicken place.

KFC stands for,

Keep. Flip. Change.

Keep the first fraction as it is.

Flip the second fraction.

Change the sign.

What about when a fraction is divided by a whole number?

Does KFC still work?

Yep. Same process.

Keep the first fraction as it is.

Flip the *number.*

Change the sign.

How do you flip a number?

Write it as a fraction by putting it over 1.

Then flip it.

You'll get 1 over that number.

That's it.

Then go ahead and multiply straight across.

Worked Example 1

Calculate the following:

a $\frac{2}{7} \times \frac{1}{5}$ b $\frac{3}{5} \times \frac{9}{10}$ c $\frac{2}{11} \times \frac{8}{9}$

Explanation	Solution
a Write the fractions.	$\frac{2}{7} \times \frac{1}{5}$
Multiply the numerators and the denominators straight across.	$= \frac{2 \times 1}{7 \times 5}$
	$= \frac{2}{35}$
b Write the fractions.	$\frac{3}{5} \times \frac{9}{10}$
Multiply the numerators and the denominators straight across.	$= \frac{3 \times 9}{5 \times 10}$
	$= \frac{27}{50}$
c Write the fractions.	$\frac{2}{11} \times \frac{8}{9}$
Multiply the numerators and the denominators straight across.	$= \frac{2 \times 8}{11 \times 9}$
	$= \frac{16}{99}$

Worked Example 2

Calculate the following:

a $4 \times \frac{3}{7}$ b $5 \times \frac{2}{9}$ c $-7 \times \frac{4}{11}$

Explanation **Solution**

a Write the fractions.

$$4 \times \frac{3}{7}$$

Multiply the whole number by the numerator, leave the denominator as is.

$$= \frac{4 \times 3}{7}$$

$$= \frac{12}{7}$$

b Write the fractions.

$$5 \times \frac{2}{9}$$

Multiply the whole number by the numerator, leave the denominator as is.

$$= \frac{5 \times 2}{9}$$

$$= \frac{10}{9}$$

c Write the fractions.

$$-7 \times \frac{4}{11}$$

Multiply the whole number by the numerator, leave the denominator as is.

$$= \frac{-7 \times 4}{11}$$

$$= -\frac{28}{11}$$

Worked Example 3

Calculate the following:

a $\mathbf{\frac{3}{5} \div \frac{2}{7}}$

b $\mathbf{\frac{7}{8} \div \frac{2}{9}}$

c $\mathbf{-\frac{4}{9} \div \frac{5}{11}}$

Explanation	Solution
a Write the fractions.	$\frac{3}{5} \div \frac{2}{7}$
Flip the second fraction and change the sign to multiplication.	$= \frac{3}{5} \times \frac{7}{2}$
Multiply the numerators and the denominators straight across.	$= \frac{21}{10}$
b Write the fractions.	$\frac{7}{8} \div \frac{2}{9}$
Flip the second fraction and change the sign to multiplication.	$= \frac{7}{8} \times \frac{9}{2}$
Multiply the numerators and the denominators straight across.	$= \frac{63}{16}$
c Write the fractions.	$-\frac{4}{9} \div \frac{5}{11}$
Flip the second fraction and change the sign to multiplication.	$= -\frac{4}{9} \times \frac{11}{5}$
Multiply the numerators and the denominators straight across.	$= -\frac{44}{45}$

Worked Example 4

Calculate the following:

a $9 \div \frac{2}{5}$ b $\frac{11}{15} \div -3$

Explanation	**Solution**
a Write the fractions.	$9 \div \frac{2}{5}$
Flip the second fraction and change the sign to multiplication.	$= 9 \times \frac{5}{2}$
Multiply the whole number by the numerator, leave the denominator as is.	$= \frac{45}{10}$
b Write the fractions.	$\frac{11}{15} \div -3$
Rewrite the whole number over 1. (optional)	$\frac{11}{15} \div \frac{-3}{1}$
Flip the whole number and change the sign to multiplication.	$= \frac{11}{15} \times \frac{1}{-3}$
Multiply the numerators and the denominators straight across.	$= -\frac{11}{45}$

When you have a fraction over another fraction,
it means divide the top fraction by the bottom fraction.

Worked Example 5

Calculate the following:

a $\dfrac{\frac{1}{2}}{\frac{4}{3}}$

b $\dfrac{\frac{2}{5}}{3}$

Explanation	**Solution**
a Write the fractions.	$\dfrac{\frac{1}{2}}{\frac{4}{3}}$
Flip the bottom fraction and multiply by the top fraction.	$= \frac{1}{2} \times \frac{3}{4}$
Multiply the numerators and the denominators straight across.	$= \frac{3}{8}$
b Write the fractions.	$\dfrac{\frac{2}{5}}{3}$
Rewrite whole number over 1.	$= \dfrac{\frac{2}{5}}{\frac{3}{1}}$
Flip the bottom fraction and multiply by the top fraction.	$= \frac{2}{5} \times \frac{1}{3}$
Multiply the numerators and the denominators straight across.	$= \frac{2}{15}$

Multiplying and Dividing Fractions: Keep Numbers Small and Friendly

Alright, the examples we just did.

Pretty straightforward.

But what if the fractions has big numbers?

One — you don't wanna be multiplying big numbers.

Two — you don't wanna be simplifying fractions with even bigger numbers.

Nobody wants that.

So, here's what to do.

Remember how you simplified fractions?

You found the Greatest Common Factor (GCF) and cancelled it out.

Same idea here.

Before you multiply, you wanna find the GCF.

Or GCFs.

Sometimes there's more than one.

Then cancel each of them out.

Doing this gives simpler fractions.

With smaller numbers.

Which makes multiplying across easier.

and also makes simplifying the final answer a whole lot easier.

And that's exactly what we want.

The GCF doesn't have to be in the same fraction.

It can be in any fraction.

In any numerator or any denominator.

And it can even be canceled out diagonally.

As long as you're multiplying, and the same number is in a numerator and a denominator.

In any of the fractions.

You can cancel it out.

Okay, what if you're dividing fractions?

Can you just cancel out?

Good question.

You can't cross-cancel when dividing fractions.

First do the flipping thing and change the sign to multiplication.

Only then cancel out any GCFs.

So, in a nutshell, when multiplying fractions,

1) Find the GCFs.

2) Cancel them out (even diagonally)

3) Multiply what's left

4) Simplify further if needed

Oh, this is important.

You can't cross-cancel the GCF when you're adding or subtracting fractions.

That's a big no-no.

You can only do that when you're multiplying.

OK?

So, make sure you don't forget that.

Worked Example 6

Simplify the following:

a $\frac{8}{9} \times \frac{6}{11}$

b $\frac{4}{7} \times \frac{5}{12}$

Explanation	**Solution**
a Write the fractions.	$\frac{8}{9} \times \frac{6}{11}$
Break down the numbers with common factors.	$= \frac{8}{3 \times 3} \times \frac{3 \times 2}{11}$
Cancel out the common factor diagonally.	$= \frac{8}{3} \times \frac{2}{11}$
Multiply the numerators and the denominators straight across.	$= \frac{16}{33}$
b Write the fractions.	$\frac{4}{7} \times \frac{5}{12}$
Break down the numbers with common factors.	$= \frac{4 \times 1}{7} \times \frac{5}{4 \times 3}$
Cancel out the common factor diagonally.	$= \frac{1}{7} \times \frac{5}{3}$
Multiply the numerators and the denominators straight across.	$= \frac{5}{21}$

Worked Example 7

Simplify the following:

a $\frac{9}{10} \times \frac{2}{15}$ b $\frac{21}{80} \times \frac{4}{15}$

Explanation		Solution
a	Write the fractions.	$\frac{9}{10} \times \frac{2}{15}$
	Break down the numbers with common factors.	$= \frac{3 \times 3}{2 \times 5} \times \frac{2 \times 1}{3 \times 5}$
	Cancel out the common factor diagonally.	$= \frac{3}{5} \times \frac{1}{5}$
	Multiply the numerators and denominators together.	$= \frac{3}{25}$
b	Write the fractions.	$\frac{21}{80} \times \frac{4}{15}$
	Break down the numbers with common factors.	$= \frac{3 \times 7}{4 \times 20} \times \frac{4 \times 1}{3 \times 5}$
	Cancel out the common factor diagonally.	$= \frac{7}{20} \times \frac{1}{5}$
	Multiply the numerators and the denominators straight across.	$= \frac{7}{100}$

Multiplying numerators and denominators straight across when dividing. ⚠

Worked Example 8

Simplify the following:

a $\frac{5}{6} \times 24$ b $\frac{3}{4} \div \frac{1}{8}$

Explanation	**Solution**
a Write the fractions.	$\frac{5}{6} \times 24$
Write the whole number over 1.	$= \frac{5}{6} \times \frac{24}{1}$
Break down the numbers with common factors.	$= \frac{5}{6} \times \frac{6 \times 4}{1}$
Cancel out the common factor diagonally.	$= \frac{5}{1} \times \frac{4}{1}$
Multiply the numerators and the denominators straight across and simplify.	$= \frac{20}{1}$
	$= 20$
b Write the fractions.	$\frac{3}{4} \div \frac{1}{8}$
Flip the second fraction and change the sign to multiplication.	$\frac{3}{4} \times \frac{8}{1}$
Break down the numbers with common factors.	$= \frac{3}{4} \times \frac{4 \times 2}{1}$
Cancel out the common factor diagonally.	$= \frac{3}{1} \times \frac{2}{1}$
Multiply the numerators and the denominators straight across and simplify.	$= \frac{6}{1}$
	$= 6$

Worked Example 9

Simplify $\left(2 \times \frac{1}{3}\right) \div \left(3 \times \frac{6}{5}\right)$

Explanation	**Solution**
Write the fractions.	$\left(2 \times \frac{1}{3}\right) \div \left(3 \times \frac{6}{5}\right)$
Calculate inside each brackets first.	$= \frac{2}{3} \div \frac{18}{5}$
Flip the second fraction and change the sign to multiplication.	$= \frac{2}{3} \times \frac{5}{18}$
Break down the numbers with common factors.	$= \frac{2}{3} \times \frac{5}{2 \times 9}$
Cancel out the common factor diagonally.	$= \frac{1}{3} \times \frac{5}{9}$
Multiply the numerators and the denominators straight across.	$= \frac{5}{27}$

When simplifying fractions with multiple operations,
follow the order of operations - BOMDAS
Brackets first, then Multiply and Divide before Add and Subtract.

Worked Example 10

Simplify $\left(\frac{1}{2}\div\frac{9}{4}\right)\times\left(11-\frac{4}{3}\right)$

Explanation	**Solution**
Write the fractions.	$\left(\frac{1}{2}\div\frac{9}{4}\right)\times\left(11-\frac{4}{3}\right)$
Change the division to multiplication in the first bracket. Rewrite the second bracket.	$=\left(\frac{1}{2}\times\frac{4}{9}\right)\times\left(\frac{11}{1}-\frac{4}{3}\right)$
Cancel out the common factor in the first bracket. Combine the fractions in the second bracket.	$=\left(\frac{1}{1}\times\frac{2}{9}\right)\times\left(\frac{33-4}{3}\right)$
Work out each bracket separately.	$=\frac{2}{9}\times\frac{29}{3}$
Multiply the numerators and the denominators straight across.	$=\frac{58}{27}$

Forgetting to flip the second fraction when dividing. Or not changing division to multiplication sign after flipping. ⚠

Your Turn To Try

1 **When simplified $\frac{2}{5} \times \frac{3}{4}$?**

A $\frac{5}{9}$

B $\frac{3}{10}$

C $\frac{1}{4}$

D $\frac{20}{23}$

2 **$\frac{5}{6} \div \frac{10}{9}$ is equal to which of the following?**

A $\frac{3}{5}$

B $\frac{5}{9}$

C $\frac{25}{27}$

D $\frac{3}{4}$

3 **$6 \div \frac{2}{5}$ when simplified is?**

A $\frac{3}{5}$

B $\frac{12}{5}$

C 15

D $\frac{7}{15}$

4 **When is the result of $\frac{9}{14} \times \frac{7}{3}$?**

A $\frac{16}{17}$

B $\frac{3}{2}$

C $\frac{27}{56}$

D $\frac{21}{14}$

5 **What is $\frac{5}{8} \div 4$?**

A $\frac{5}{32}$

B $\frac{5}{4}$

C $\frac{5}{12}$

D $\frac{5}{2}$

Question	1	2	3	4	5
Answer	B	D	C	B	A

Time To Practice

1 Simplify the following.

a $\frac{2}{9}\times\frac{1}{5}$ b $\frac{1}{5}\times\frac{3}{2}$ c $\frac{7}{8}\times\frac{9}{11}$

2 Work out the following.

a $\frac{2}{3}\times 4$ b $\frac{2}{3}\div 4$ c $4\div\frac{2}{3}$

3 Multiply and simplify.

a $\frac{5}{6}\times\frac{18}{25}$ b $-\frac{7}{9}\times\frac{45}{49}$ c $\frac{3}{40}\times\frac{5}{33}$

4 Divide the following.

a $\frac{1}{6}\div\frac{4}{5}$ b $\frac{2}{3}\div\frac{7}{8}$ c $\frac{4}{9}\div\frac{5}{8}$

5 Evaluate each fraction.

a $\left(\frac{2}{3}\div\frac{1}{4}\right)\div 2$ b $\frac{5}{9}\div\left(\frac{5}{6}\div\frac{2}{3}\right)$ c $\left(\frac{1}{2}\div\frac{45}{40}\right)\div\frac{40}{25}$

6 Simplify and solve.

a $\left(\frac{1}{2}-\frac{2}{3}\right)\times\frac{3}{4}$ b $\frac{3}{5}\div\left(\frac{1}{4}+\frac{1}{6}\right)$ c $\left(\frac{5}{8}\times\frac{2}{3}\right)-\frac{1}{4}$

7 Simplify each fraction to its lowest terms.

a $\left(\frac{12}{25}\div\frac{8}{5}\right)\times\frac{7}{10}$ b $\left(\frac{55}{30}\div\frac{40}{30}\right)\times\frac{10}{11}$ c $\frac{5}{2}\times\left(\frac{160}{30}\div\frac{2}{15}\right)$

Challenge Yourself

1 Perform the operations in order and simplify your final answer

a $\left(\frac{8}{9} \div \frac{2}{3}\right) \div \left(\frac{3}{4} \div \frac{1}{2}\right)$

b $\left(\frac{11}{12} \div \frac{5}{6}\right) \div \left(\frac{3}{14} \div \frac{1}{4}\right)$

2 Simplify the following.

a $\left(\frac{5}{8} \div \frac{1}{4}\right) \times \left(\frac{3}{5} - \frac{1}{10}\right)$

b $\frac{4}{5} \times \left(\frac{2}{3} + \frac{1}{6}\right) \div \frac{3}{4}$

3 Calculate the result of the following.

a $\left(\frac{18}{25} \div \frac{12}{35}\right) + \left(\frac{16}{45} \times \frac{27}{32}\right)$

b $\left(\frac{24}{35} \times \frac{15}{28}\right) - \left(\frac{3}{14} \div \frac{21}{2}\right)$

Decimals to Fractions: A Quick Switch For When You Need It

There's one more basic skill you need to be good at.

Turning decimals into fractions.

This skill's gonna be useful later on.

Especially when you get into higher-level math.

Because sometimes, to get to the answer, it's easier to use decimals.

Other times, it's easier to use fractions.

And if you don't know how to switch between the two, you'll end up stuck.

That's why we're covering this.

Nothing too complicated.

We'll focus on normal decimals, not the ones that go on forever.

Here's how you turn decimals into fractions,

1) Write the decimal as a fraction over 1 $\left(\frac{decimal}{1}\right)$
2) Multiply by $\frac{10}{10}$, $\frac{100}{100}$ or $\frac{1000}{1000}$. Match the number of zeros to the number of decimal places
3) Simplify if needed

Quick reminder.

Multiplying a decimal by 10, 100, or 1000 moves the decimal point to the right.

One place for each zero.

If there's a number in front of the decimal.

Follow the exact same 3 steps.

It'll still work.

Worked Example 1

Convert the following decimals to fractions, in their simplest forms.

a $\mathbf{0.1}$ b $\mathbf{0.9}$ c $\mathbf{0.27}$

Explanation	**Solution**
a Write the decimal.	0.1
Rewrite decimal as a fraction over 1, and multiply it by $\frac{10}{10}$.	$= \frac{0.1}{1} \times \frac{10}{10}$
Multiply the numerators and the denominators straight across.	$= \frac{1}{10}$
b Write the decimal.	0.9
Rewrite decimal as a fraction over 1, and multiply it by $\frac{10}{10}$.	$= \frac{0.9}{1} \times \frac{10}{10}$
Multiply the numerators and the denominators straight across.	$= \frac{9}{10}$
c Write the decimal.	0.27
Rewrite decimal as a fraction over 1, and multiply it by $\frac{100}{100}$.	$= \frac{0.27}{1} \times \frac{100}{100}$
Multiply the numerators and the denominators straight across.	$= \frac{27}{100}$

Worked Example 2

Convert the following decimals to fractions, in their simplest forms.

a $\mathbf{0.16}$ b $\mathbf{0.45}$

Explanation	**Solution**
a Write the decimal.	0.16
Rewrite decimal as a fraction over 1, and multiply it by $\frac{100}{100}$.	$= \frac{0.16}{1} \times \frac{100}{100}$
Multiply the numerators and the denominators straight across.	$= \frac{16}{100}$
Simplify.	$= \frac{4 \times 4}{4 \times 25} = \frac{4}{25}$
b Write the decimal.	0.45
Rewrite decimal as a fraction over 1, and multiply it by $\frac{100}{100}$.	$= \frac{0.45}{1} \times \frac{100}{100}$
Multiply the numerators and the denominators straight across.	$= \frac{45}{100}$
Simplify.	$= \frac{5 \times 9}{5 \times 20} = \frac{9}{20}$

Worked Example 3

Convert the following decimals to fractions, in their simplest forms.

a 1.5 b 1.12

Explanation	Solution
a Write the decimal.	1.5
Rewrite decimal as a fraction over 1, and multiply it by $\frac{10}{10}$.	$= \frac{1.5}{1} \times \frac{10}{10}$
Multiply the numerators and the denominators straight across.	$= \frac{15}{10}$
Simplify.	$= \frac{3}{2}$
b Write the decimal.	1.12
Rewrite decimal as a fraction over 1, and multiply it by $\frac{100}{100}$.	$= \frac{1.12}{1} \times \frac{100}{100}$
Multiply the numerators and the denominators straight across.	$= \frac{112}{100}$
Simplify.	$= \frac{4 \times 28}{4 \times 25}$
	$= \frac{28}{25}$

Your Turn To Try

1 **What is 0.45 as a fraction?**

A $\frac{2}{3}$

B $\frac{1}{4}$

C $\frac{9}{20}$

D $\frac{4}{5}$

2 **How do you write 0.66 as a fraction?**

A $\frac{3}{5}$

B $\frac{33}{50}$

C $\frac{5}{3}$

D $\frac{2}{3}$

3 **Which fraction does NOT equal 0.7?**

A $\frac{7}{10}$

B $\frac{70}{100}$

C $\frac{21}{30}$

D $\frac{3}{4}$

4 **What is the fraction form of 1.5?**

A $\frac{5}{3}$

B $\frac{3}{2}$

C $\frac{4}{3}$

D $\frac{7}{5}$

5 **What fraction equals 2.25?**

A $\frac{9}{4}$

B $\frac{7}{4}$

C $\frac{5}{4}$

D $\frac{11}{4}$

Question	1	2	3	4	5
Answer	C	B	D	B	A

Time To Practice

1 Convert the following decimals to fractions.

a 0.3	b 0.7	c 0.8
d 0.12	e 0.13	f 0.22
g 0.31	h 0.42	i 0.55
j 0.62	k 0.78	l 0.82
m 0.88	n 0.92	o 0.95

2 Write the following decimals as fractions, in their simplest form.

a 1.3	b 1.4	c 1.6
d 1.9	e 2.5	f 3.9
g 1.46	h 1.72	i 1.85

Challenge Yourself

1 Write the following decimals as fractions, in their simplest form.

a 0.03	b 0.04	c 0.005
d 0.123	e 0.402	f 0.888

1. Answers

Divisibility Rules

Time to Practice

1	Yes		2	Yes		3	Yes	
4	No		5	Yes		6	Yes	
7	Yes		8	No		9	No	
10	No		11	No		12	Yes	
13	a	2, 3, 4, 5, 6, 8, 10	b	2, 3, 4, 6		c	2, 3, 6, 9	
	d	2, 3, 4, 5, 6, 8, 10	e	3, 7		f	2, 3, 6, 7, 9	
	g	2, 7	h	2, 3, 4, 5, 6, 8, 10		i	2, 4, 8	
	j	2, 3, 4, 6, 7, 8, 9	k	None		l	2, 3, 4, 5, 6, 7, 8, 10	

Challenge Yourself

1	1 or 4	2	8	3	3
4	9	5	5		

Simplifying Fractions

Time to Practice

1 a $\frac{1}{9}$ b $\frac{1}{7}$ c $\frac{1}{3}$

2 a $\frac{1}{7}$ b $\frac{1}{8}$ c $\frac{1}{11}$

3 a $\frac{2}{3}$ b $\frac{2}{5}$ c $\frac{2}{9}$

4 a $\frac{3}{5}$ b $\frac{2}{5}$ c $\frac{2}{3}$

5 a $\frac{4}{9}$ b $\frac{6}{7}$ c $\frac{7}{8}$

6 a $\frac{5}{8}$ b $\frac{5}{9}$ c $\frac{5}{6}$

7 a $\frac{8}{11}$ b $\frac{7}{9}$ c $\frac{11}{12}$

Challenge Yourself

1 a $\frac{2}{3}$ b $\frac{3}{4}$ c $\frac{4}{5}$

2 a $\frac{5}{6}$ b $\frac{5}{7}$ c $\frac{10}{11}$

Adding and Subtracting Fractions

Time to Practice

1	a $\frac{3}{4}$	b $\frac{1}{4}$	c 1
2	a $\frac{1}{6}$	b $\frac{1}{10}$	c $-\frac{1}{6}$
3	a $\frac{7}{8}$	b $\frac{61}{70}$	c $\frac{39}{88}$
4	a $\frac{7}{4}$	b $-\frac{5}{6}$	c $\frac{37}{10}$
5	a $\frac{7}{8}$	b $\frac{1}{2}$	c $\frac{17}{10}$
6	a $\frac{11}{20}$	b $-\frac{7}{36}$	c $\frac{17}{24}$
7	a $-\frac{47}{30}$	b $-\frac{1}{4}$	c $-\frac{4}{5}$

Challenge Yourself

1	a 1	b $\frac{19}{20}$
2	a $-\frac{1}{9}$	b $-\frac{13}{30}$
3	a $\frac{1}{27}$	b $-\frac{9}{40}$

Multiplying and Dividing Fractions

Time to Practice

1 a $\frac{2}{45}$ b $\frac{3}{10}$ c $\frac{63}{88}$

2 a $\frac{8}{3}$ b $\frac{1}{6}$ c 6

3 a $\frac{3}{5}$ b $-\frac{5}{7}$ c $\frac{1}{88}$

4 a $\frac{5}{24}$ b $\frac{16}{21}$ c $\frac{32}{45}$

5 a $\frac{4}{3}$ b $\frac{4}{9}$ c $\frac{5}{18}$

6 a $-\frac{1}{8}$ b $\frac{36}{25}$ c $\frac{1}{6}$

7 a $\frac{21}{100}$ b $\frac{5}{4}$ c 100

Challenge Yourself

1 a $\frac{8}{9}$ b $\frac{77}{60}$

2 a $\frac{5}{4}$ b $\frac{8}{9}$

3 a $\frac{12}{5}$ b $\frac{17}{49}$

Decimals to Fractions

Time to Practice

1 a $\frac{3}{10}$ b $\frac{7}{10}$ c $\frac{4}{5}$

d $\frac{3}{25}$ e $\frac{13}{100}$ f $\frac{11}{50}$

g $\frac{31}{100}$ h $\frac{21}{50}$ i $\frac{11}{20}$

j $\frac{31}{50}$ k $\frac{39}{50}$ l $\frac{41}{50}$

m $\frac{22}{25}$ n $\frac{23}{25}$ o $\frac{19}{20}$

2 a $\frac{13}{10}$ b $\frac{7}{5}$ c $\frac{8}{5}$

d $\frac{19}{10}$ e $\frac{5}{2}$ f $\frac{39}{10}$

g $\frac{73}{50}$ h $\frac{43}{25}$ i $\frac{37}{20}$

Challenge Yourself

1 a $\frac{3}{100}$ b $\frac{1}{25}$ c $\frac{1}{200}$

d $\frac{123}{1000}$ e $\frac{201}{500}$ f $\frac{111}{125}$

Exponents
Powers

Exponents: Math's Way of Saying 'Times This by Itself... A Lot'

We've got the basics out of the way.

Now you'll need to get good at exponents.

Or Powers.

Some places call them exponents.

Some call them powers.

They mean the same thing.

So, what's an exponent?

It's basically the number of times you multiply something by itself.

And it's written in a specific way.

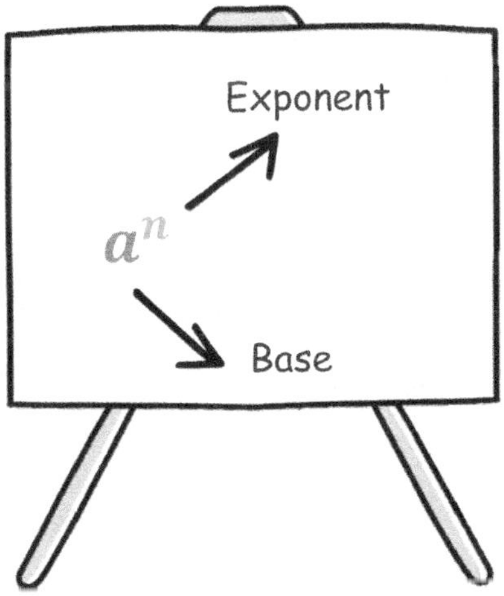

There's the base.

The bigger number at the bottom.

That's what you're multiplying.

And there's the exponent.

The smaller number up on the top right.

That tells you how many times to multiply the base by itself.

This way of writing is called *exponential form.*

The way you say this is,

"a raised to the power of n" or "a to the power of n".

More generally it's,

(the base) to the power of (the exponent).

Have a look at these examples.

$$2^5 = 2 \times 2 \times 2 \times 2 \times 2 = \mathbf{32}$$

$$(-3)^3 = (-3) \times (-3) \times (-3) = \mathbf{-27}$$

$$\left(\frac{1}{2}\right)^4 = \left(\frac{1}{2}\right) \times \left(\frac{1}{2}\right) \times \left(\frac{1}{2}\right) \times \left(\frac{1}{2}\right) = \mathbf{\frac{1}{16}}$$

$$a^7 = a \times a \times a \times a \times a \times a \times a$$

See how the exponential form is way easier to write?

That's the whole point of exponents.

It's a shorter way of writing all that repeated multiplication.

Like, who wants to write 'two times two times two times two' every time?

Or any number multiplied by itself over and over again.

Nobody.

Alright.

That's the basics.

Now let's look at how to actually work with them.

Exponent Laws: 9 Rules You Can't Live Without (Mathematically Speaking)

When working with exponents, there are specific rules to follow.

These rules — sometimes called laws — tell you how to work with exponents.

There are 9 of them.

All of which.

You. Have. To. Memorize.

Seriously.

If you want any hope of doing well in Algebra, or beyond.

You've got to know these.

They're never given to you in tests.

And if you don't know when to use them?

Things get confusing.

Fast.

A lot of people struggle with Algebra.

And it's usually because they didn't learn the laws properly.

Or didn't bother memorizing them.

So, here's what we're gonna do.

We'll go through each law.

Nice and slow.

One at a time.

Starting with the first three.

The rest build from there.

As we go, make sure you understand what's happening.

And try to memorize each rule.

With every new law introduced, the questions get harder.

To solve them, you'll need to use the new law along with the ones already covered.

Eventually, towards the end,

you'll see crazy looking expressions.

And to simplify them, you'll need to use a mix of the laws.

And if you forget any of them.

Well, you just won't be able to do it.

Alright, here are the first three laws.

Exponent Laws

Law 1	Anything to the power of 0 equals 1	$a^0 = 1$
Law 2	Same Base Multiplication. Keep base, add powers	$a^m \times a^n = a^{m+n}$
Law 3	Same Base Division. Keep base, subtract powers	$a^m \div a^n = a^{m-n}$

Just a heads-up.

These laws only apply when the bases are the same.

If the bases are different, the law does not apply.

You'll also see letters like a and b for bases, and m and n for exponents.

These could be any number or letter — doesn't matter.

The math stays the same.

Cool?

Let's get into some examples.

Worked Example 1

Simplify each of the following.

a a^0 b $(2b^3)^0$ c $-4(a^2b^5)^0$ d $7x^0 - 4(2y)^0$

	Explanation	Solution
a	Write the expression.	x^0
	Apply Law 1. (Anything to the power of 0 equals 1).	$= 1$
b	Write the expression.	$(2b^3)^0$
	Apply Law 1.	$= 1$
c	Write the expression.	$-4(a^2b^5)^0$
	Apply Law 1.	$= -4 \times 1$
	Simplify.	$= -4$
d	Write the expression.	$7x^0 - 4(2y)^0$
	Apply Law 1.	$= 7 \times 1 - 4 \times 1$
	Simplify.	$= 7 - 4$ $= 3$

When there's no exponent, it means the power is 1.
e.g. $a = a^1$ and $5a = 5^1 \times a^1$.

Worked Example 2

Simplify, using the exponent law for multiplication.

a $4^5 \times 4^2$ b $a^7 \times a^1$ c $2b \times 9b^5$ d $m^2 \times m^3 \times m$

Explanation		Solution
a	Write the expression.	$4^5 \times 4^2$
	Apply Law 2: (Same base. Add the powers.)	$= 4^{5+2}$
	Simplify.	$= 4^7$
b	Write the expression.	$a^7 \times a^1$
	Apply Law 2.	$= a^{7+1}$
	Simplify.	$= a^8$
c	Write the expression.	$2b \times 9b^5$
	Multiply the numbers. Apply Law 2.	$= 18b^{1+5}$
	Simplify.	$= 18b^6$
d	Write the expression.	$m^2 \times m^3 \times m$
	Apply Law 2.	$= m^{2+3+1}$
	Simplify.	$= m^6$

If there are numbers in front of the same bases, just multiply or divide them as you normally would.

Worked Example 3

Simplify, using the exponent law for division:

a $3^9 \div 3^2$ b $a^7 \div a^1$ c $\dfrac{8b^7}{6b^2}$ d $\dfrac{m^9 \times m^2p^2}{m^8}$

Explanation	Solution
a Write the expression.	$3^9 \div 3^2$
Apply Law 3. (Same base. Subtract the powers.)	$= 3^{9-2}$
Simplify.	$= 3^7$
b Write the expression.	$a^7 \div a^1$
Apply Law 3.	$= a^{7-1}$
Simplify.	$= a^6$
c Write the expression.	$\dfrac{8b^7}{6b^2}$
Divide the numbers. Apply Law 3.	$= \dfrac{4}{3}b^{7-2}$
Simplify.	$= \dfrac{4}{3}b^5 = \dfrac{4b^5}{3}$
d Write the expression.	$\dfrac{m^9 \times m^2p^2}{m^8}$
Apply Law 2 for the numerator.	$= \dfrac{m^{11}p^2}{m^8}$
Apply Law 3.	$= m^{11-8}p^2$
	$= m^3p^2$

Worked Example 4

Simplify each of the following:

a $\dfrac{6a^8b^4 \times 2a^7b^6}{9a^5b^2}$ b $\dfrac{3a^6b^5}{a^5b^3} \times \dfrac{a^4}{b^2}$

Explanation	Solution
a Write the expression.	$\dfrac{6a^8b^4 \times 2a^7b^6}{9a^5b^2}$
Apply Law 2 to numerator.	$= \dfrac{12a^{15}b^{10}}{9a^5b^2}$
Simplify numbers. Apply Law 3.	$= \dfrac{4a^{15-5}b^{10-2}}{3}$
Simplify.	$= \dfrac{4a^{10}b^8}{3}$
b Write the expression.	$\dfrac{3a^6b^5}{a^5b^3} \times \dfrac{a^4}{b^2}$
Apply Law 2 and multiply the numerators and denominators.	$= \dfrac{3a^{10}b^5}{a^5b^5}$
Apply Law 3.	$= \dfrac{3a^{10-5}b^{5-5}}{1}$ $= 3a^5b^0$
Simplify.	$= 3a^5$

When multiplying or dividing different bases, apply the exponent laws to each base separately.

Worked Example 5

Simplify $\dfrac{7a^5b^3}{5a^6b^2} \div \dfrac{7b^3a^2}{5b^5a^4}$

Explanation	Solution
Write the expression	$\dfrac{7a^5b^3}{5a^6b^2} \div \dfrac{7b^3a^2}{5b^5a^4}$
Flip the second fraction and change the sign to multiplication.	$= \dfrac{7a^5b^3}{5a^6b^2} \times \dfrac{5b^5a^4}{7b^3a^2}$
Cancel out common number factors.	$= \dfrac{a^5b^3}{a^6b^2} \times \dfrac{b^5a^4}{b^3a^2}$
Apply Law 2. Multiply across.	$= \dfrac{a^9b^8}{a^8b^5}$
Apply Law 3.	$= a^{9-8}b^{8-5}$
Simplify.	$= ab^3$

Not continuing to apply the laws repeatedly until the expression is fully simplified.

Your Turn To Try

1 **What is the result of $x^3 \times x^4$?**

A x^{12}

B x^9

C x^1

D x^7

2 **What is $a^5 \div a^2$ equivalent to?**

A a^{10}

B a^3

C a^7

D a^2

3 **What do you get when you multiply $3y^2 \times 2y^5$**

A $6y^7$

B $6y^3$

C $5y^7$

D $5y^3$

4 **What is the result of $\frac{p^4 \times p^2}{p^3}$**

A p^{12}

B p^9

C p^3

D p^7

5 **What is $x^0 \times x^4$**

A 1

B x^4

C 0

D x^{-4}

Question	1	2	3	4	5
Answer	D	B	A	C	B

Time To Practice

1 Simplify the following.

a $2^3 \times 2^2$

b $3^2 \times 3$

c $5^0 \times 5^3$

d $x^6 \times x^0$

e $y^3 \times y^2$

f $z^9 \times z^7$

g $a^3 \times a^4 \times a^6$

h $b^3 \times b^5 \times b^2$

i $c^2 \times c^8 \times c^9$

j $5 \times (2x^2)^0$

k $(3x^2y)^0 \times 7$

l $3(2x^2y^3)^0 - 4$

m $6a^4 \times 2a^2$

n $10b^6 \times 2b^3$

o $5c^0 \times 7c^5 \times c^2$

2 Write the following expressions in simplest index form.

a $7^6 \div 7^4$

b $4^{12} \div 4^9$

c $10^{17} \div 10^{14}$

d $m^9 \div m^2$

e $n^{20} \div n^5$

f $p^{15} \div p^3$

g $\frac{5^5}{5^4} \times 5$

h $\frac{3^8}{3^6} \times 3^2$

i $\frac{2^8}{2^5} \times 2^3$

j $\frac{10x^5}{2x^2} \times 3x^3$

k $\frac{14y^8 \times y^0}{7y^4}$

l $\frac{50z^6 \times z^6}{5z^2}$

m $4a^3 \times 3a^4 \div 2a^2$

n $12b^9 \times 3b^2 \div 4b^0$

o $16c^7 \times 2c^3 \div 8c^2$

Challenge Yourself

1 Simplify each of the following.

a $\dfrac{12u^{11}v^9 \times 3u^2v^3}{18u^6v^5}$

b $\dfrac{5e^6f^4 \times 8e^4f^3}{10ef^5}$

2 Write the following in simplest index notation.

a $\dfrac{-3ab^3}{ab} \div \dfrac{b}{a^4b^2}$

b $\dfrac{16a^7b^6}{6a^3b^3} \div \dfrac{8a^6b^3}{27a^3b^2}$

3 Simplify the following.

a $\dfrac{8x^2y \times 3x^5y^2}{6x^2y^2} \times \dfrac{7x^5y^2}{16x^3y}$

b $\dfrac{16x^4y^8}{27x^4y^9} \div \dfrac{x^6y^2}{x^5y^5z^3 \times xyz}$

Power of a Power: Ready for Round Two?

Let's check out the next set — Laws 4, 5, and 6.

Use these when you see,

- A power raised to another power
- A bunch of terms inside brackets, raised to a power
- A fraction raised to a power

Sometimes all three show up in one expression.

Just remember this.

When you've got brackets with a power outside.

These are the laws to use.

Exponent Laws

Law 4	Exponential to a Power: Keep base, multiply powers	$(a^m)^n = a^{mn}$
Law 5	Multiplication to a Power: Raise each term inside to power outside	$(ab)^n = a^n b^n$
Law 6	Fraction to a Power: Raise numerator and denominator to power	$\left(\frac{a}{b}\right)^n = \frac{a^n}{b^n}$

Power of a Number

Working with powers, you'll need to know what certain numbers equal when raised to a power.

Here are some you'll see a lot.

Try to memorize these too.

n	**1**	**2**	**3**	**4**
2^n	2	4	8	16
3^n	3	9	27	81
4^n	4	16	64	256
5^n	5	25	125	625

Another thing before we get into the examples.

As we go through laws 4, 5 and 6, and the ones coming up.
The questions will start getting trickier.

Now.
There's no single "right" way to answer them.
Sometimes, there'll be more than one way.

For instance.
You can combine fractions first, then simplify.
Or, you can simplify first, then combine the fractions.
Both ways work.

The main thing is to apply the laws properly.
Do that, and you'll always get the correct answer.
So just keep that in mind.

Worked Example 1

Simplify using the laws involving brackets.

a $(4^3)^5$ b $(a^5)^2$ c $4(b^2)^3$ d $a(b^8)^2$

Explanation	Solution
a Write the expression.	$(4^3)^5$
Apply Law 4. (Keep base, multiply powers)	$= 4^{3\times 5}$
Simplify.	$= 4^{15}$
b Write the expression.	$(a^5)^2$
Apply Law 4.	$= a^{5\times 2}$
Simplify.	$= a^{10}$
c Write the expression.	$4(b^2)^3$
Apply Law 4. (The number 4 is outside the brackets, so it stays as it is)	$= 4b^{2\times 3}$
Simplify.	$= 4b^6$
d Write the expression.	$a(b^8)^2$
Apply Law 4. ('a' is outside the brackets so it stays as it is)	$= ab^{8\times 2}$
Simplify.	$= ab^{16}$

Worked Example 2

Simplify and evaluate the following using the laws involving brackets.

a $(2x^4)^3$ b $(-4)^3$ c $(a^2b^7)^3$

Explanation	Solution
a Write the expression.	$(2x^4)^3$
Apply Law 5. (Raise each term inside brackets to the power)	$= 2^{1\times3} \times x^{4\times3}$ $= 2^3x^{12}$
Evaluate and simplify.	$= 8x^{12}$
b Write the expression.	$(-4)^3 = (-1 \times 4)^3$
Apply Law 5.	$= (-1)^{1\times3} \times 4^{1\times3}$
Evaluate and simplify.	$= -1 \times 4^3$ $= -64$
c Write the expression.	$(a^2b^7)^3$
Apply Law 5.	$= a^{2\times3} \times b^{7\times3}$
Evaluate and simplify.	$= a^6b^{21}$

A negative number raised to an even power becomes positive, but raised to an odd power it stays negative.

Worked Example 3

Simplify and evaluate the following using the laws involving brackets.

a $(3p^2q^2)^4$　　b $5(x^2y)^3$　　c $(-2m^3x^4)^2$

Explanation		**Solution**
a	Write the expression.	$(3p^2q^2)^4$
	Apply Law 5.	$= 3^4 \times p^{2\times4} \times q^{2\times4}$
	Evaluate and simplify.	$= 81p^8q^8$
b	Write the expression.	$5(x^2y)^3$
	Apply Law 5.	$= 5 \times x^{2\times3} \times y^{1\times3}$
	Evaluate and simplify.	$= 5x^6y^3$
c	Write the expression.	$(-2m^3x^4)^2$
	Apply Law 5.	$= (-2)^2 \times m^{3\times2} \times x^{4\times2}$
	Evaluate and simplify.	$= 4 \times m^6 \times x^8$
		$= 4m^6x^8$

Thinking $-(1)^2$ is the same as $(-1)^2$. They are different.

The minus sign outside the brackets doesn't get raised to the power, but the minus inside does. $-(1)^2 = -1$ but $(-1)^2 = 1$.

Worked Example 4

Simplify and evaluate each of the following:

a $2p(q^5)^2 \times p^4$

b $(5u^2v)^2 \times 3(2u^3)^2$

Explanation		Solution
a	Write the expression.	$2p(q^5)^2 \times p^4$
	Apply Law 4 to the brackets.	$= 2p \times q^{10} \times p^4$
	Apply Law 2. (Keep base, add powers)	$= 2p^{1+4}q^{10}$
	Simplify.	$= 2p^5q^{10}$
b	Write the expression.	$(5u^2v)^2 \times 3(2u^3)^2$
	Apply Law 5 to each of the brackets.	$= 5^2u^{2\times2}v^{1\times2} \times 3 \times 2^2u^{3\times2}$
	Apply Law 2.	$= 25u^4v^2 \times 12u^6$
	Simplify.	$= 300u^{4+6}v^2$
		$= 300u^{10}v^2$

Adding the exponents instead of multiplying when applying the power of a power rule.

Worked Example 5

Simplify each of the following:

a $\left(\frac{2}{3}\right)^3$

b $\left(\frac{a^6}{b^7}\right)^5$

Explanation	Solution
a Write the expression.	$\left(\frac{2}{3}\right)^3$
Apply Law 6. (Raise numerator and denominator to the power separately)	$=\frac{2^3}{3^3}$
Simplify.	$=\frac{8}{27}$
b Write the expression.	$\left(\frac{a^6}{b^7}\right)^5$
Apply Law 6.	$=\frac{a^{6\times 5}}{b^{7\times 5}}$
Simplify.	$=\frac{a^{30}}{b^{35}}$

Forgetting to raise both the numerator and denominator to the power when you have fractions inside brackets.

Worked Example 6

Simplify each of the following:

a $\left(\dfrac{x^4y^2}{z^3}\right)^5$ b $\left(\dfrac{3a^2}{5b}\right)^3$

Explanation	Solution
a Write the expression.	$\left(\dfrac{x^4y^2}{z^3}\right)^5$
Apply Law 5 and 6. (Raise each term inside brackets in both numerator and denominator to the power)	$= \dfrac{x^{4\times5} \times y^{2\times5}}{z^{3\times5}}$
Simplify.	$= \dfrac{x^{20}y^{10}}{z^{15}}$
b Write the expression.	$\left(\dfrac{3a^2}{5b}\right)^3$
Apply Law 5 and 6.	$= \dfrac{3^3 \times a^{2\times3}}{5^3 \times b^3}$
Simplify.	$= \dfrac{27a^6}{125b^3}$

When the same base appears in both top and bottom, subtract the smaller power from the bigger power.
The base stays where the bigger power is (top or bottom).

Worked Example 7

Simplify each of the following:

a $\left(\dfrac{x^3yz^2}{3xy^3z^6}\right)^3$

b $\dfrac{2x^3y^2}{xy^5} \times \left(\dfrac{x^4y}{2x^3y^2}\right)^4$

Explanation	**Solution**
a Write the expression	$\left(\dfrac{x^3yz^2}{3xy^3z^6}\right)^3$
Simplify fraction by applying Law 3.	$= \left(\dfrac{x^2}{3y^2z^4}\right)^3$
Apply Law 5.	$= \dfrac{x^6}{3^3y^6z^{12}}$
	$= \dfrac{x^6}{27y^6z^{12}}$
b Write the expression	$\dfrac{2x^3y^2}{xy^5} \times \left(\dfrac{x^4y}{2x^3y^2}\right)^4$
Simplify each fraction by applying Law 3.	$= \dfrac{2x^2}{y^3} \times \left(\dfrac{x}{2y}\right)^4$
Apply Law 5 to fraction inside brackets.	$= \dfrac{2x^2}{y^3} \times \dfrac{x^4}{2^4y^4}$
Multiply across and apply Law 2.	$= \dfrac{2x^6}{16y^7}$
Simplify.	$= \dfrac{x^6}{8y^7}$

Worked Example 8

Simplify $\left(\frac{2x^3y}{x^2y^2}\right)^4 \div \left(\frac{x^3y^3}{x^4y^3}\right)^3$

Explanation	**Solution**
Write the expression	$\left(\frac{2x^3y}{x^2y^2}\right)^4 \div \left(\frac{xy^5}{x^4y^3}\right)^3$
Simplify each fraction by applying Law 3.	$= \left(\frac{2x}{y}\right)^4 \div \left(\frac{y^2}{x^3}\right)^3$
Apply Law 5 to each bracket.	$= \frac{2^4x^4}{y^4} \div \frac{y^6}{x^9}$
Flip the second fraction. Change division sign to multiplication.	$= \frac{16x^4}{y^4} \times \frac{x^9}{y^6}$
Multiply across and apply Law 2.	$= \frac{16x^{13}}{y^{10}}$

Not applying the power to each term inside the brackets, including the numbers. ⚠

Worked Example 9

Simplify each of the following:

a $\left(\left(x^2\right)^3\right)^2$

b $\left(\left(\frac{a^2}{b}\right)^3\right)^5$

Explanation	**Solution**
a Write the expression.	$\left(\left(x^2\right)^3\right)^2$
Apply Law 4 for the inner brackets.	$=\left(x^6\right)^2$
Apply Law 4 for the outer brackets.	$=x^{12}$
OR	
Write the expression.	$\left(\left(x^2\right)^3\right)^2$
Apply Law 4 and multiply the powers.	$=x^{2\times3\times2}$
Simplify.	$=x^{12}$
b Write the expression.	$\left(\left(\frac{a^2}{b}\right)^3\right)^5$
Apply Law 4 and 6: Multiply the powers for numerator and denominator.	$=\frac{a^{2\times3\times5}}{b^{1\times3\times5}}$
Simplify.	$=\frac{a^{30}}{b^{15}}$

Worked Example 10

Simplify $\dfrac{8p^6q^2 \times (3p)^3q^5}{6p^4q} \div \dfrac{(2p)^2}{q^2}$

Explanation	Solution
Write expression	$\dfrac{8p^6q^2 \times (3p)^3q^5}{6p^4q} \div \dfrac{(2p)^2}{q^2}$
Apply Law 5 to brackets in the numerators and evaluate numbers.	$= \dfrac{8p^6q^2 \times 27p^3q^5}{6p^4q} \div \dfrac{4p^2}{q^2}$
Apply Law 2 to numerator of the first fraction. Flip the second fraction and change division to multiplication.	$= \dfrac{8 \times 27p^9q^7}{6p^4q} \times \dfrac{q^2}{4p^2}$
Apply Law 3 to the first fraction and cancel out common factors.	$= \dfrac{2 \times 27p^5q^6}{6} \times \dfrac{q^2}{p^2}$
	$= \dfrac{2 \times 3 \times 9p^5q^6}{2 \times 3} \times \dfrac{q^2}{p^2}$
Multiply across and apply Law2.	$= \dfrac{9p^5q^8}{p^2}$
Apply Law 3 and simplify.	$= 9p^3q^8$

Before multiplying big numbers, cancel out any common factors to make the numbers smaller.

Worked Example 11

Simplify $\frac{6x^4y^3}{16x^7y^6} \div \left(\frac{3x^2y}{2x^3y^2}\right)^3$

Explanation	**Solution**
Write the expression	$\frac{6x^4y^3}{16x^7y^6} \div \left(\frac{3x^2y}{2x^3y^2}\right)^3$
Simplify both fractions by applying law 3.	$= \frac{3}{8x^3y^3} \div \left(\frac{3}{2xy}\right)^3$
Apply Law 5 to bracket term.	$= \frac{3}{8x^3y^3} \div \frac{27}{8x^3y^3}$
Flip the second fraction. Change division sign to multiplication.	$= \frac{3}{8x^3y^3} \times \frac{8x^3y^3}{27}$
Cancel common factors.	$= \frac{3}{1} \times \frac{1}{27}$
Simplify.	$= \frac{3}{1} \times \frac{1}{3 \times 9}$
	$= \frac{1}{9}$

Your Turn To Try

1 **What is the simplified form of $(2ab^3)^4$?**

A $2a^4b^{12}$

B $8a^4b^{12}$

C $16a^4b^{12}$

D $16a^5b^7$

2 **When simplified $x^2y^2 \times (-3x^2y)^2$ equivalent to?**

A $9x^6y^4$

B $-9x^6y^4$

C $3x^6y^4$

D $-3x^6y^4$

3 **What is the result of $\left(\frac{3x^2y}{4z}\right)^3$?**

A $\frac{27x^6y^3}{64z^3}$

B $\frac{9x^6y^3}{12z^3}$

C $\frac{9x^5y^4}{12z^4}$

D $\frac{27x^5y^4}{64z^4}$

4 $\frac{6a^6b^5}{a^5b^3} \times \frac{a^4}{(2b)^2}$ **can be simplified to:**

A $3a^5b^4$

B $3a^5$

C $\frac{3a^5b}{2}$

D $\frac{3a^5}{2}$

5 $-\frac{3x^6y^3}{xy} \div \left(\frac{x^2y}{y^3}\right)^2$ **is equal to:**

A $3xy^6$

B $-3xy^6$

C $3x^6y$

D $-3x^6y$

Question	1	2	3	4	5
Answer	C	A	A	D	B

Time To Practice

1 Simplify each of the following.

a $(a^2)^5$

b $(b^5)^2 \times (b^2)^3$

c $(2a^6)^3$

d $(-3x^4)^3$

e $(x^2y)^3 \times (yx^3)^2$

f $(xy^2)^8 \div x^4y^5$

g $(2x^2y^3)^2 \times xy^4$

h $(2x^3y^2)^4 \times (x^2y)^3$

i $(2x^2y^3)^4 \div (4x^4)^2$

j $(2m^3n)^5 \div (8mn^2)^2$

k $(2p)^5q^6 \div 9p^4q^3$

l $r^6s^2 \times 2(3r)^3s^5$

2 Simplify each of the following.

a $\left(\frac{a}{b^2}\right)^2$

b $\left(\frac{a^2}{b^3}\right)^5$

c $\left(\frac{4a^3}{b^2}\right)^2$

d $\left(\frac{x^2}{y^3}\right)^4 \times \left(\frac{x^3}{y}\right)^5$

e $\left(\frac{2x^3}{y^2}\right)^2 \times \left(\frac{x^3}{2y}\right)^4$

f $\left(\frac{3x^2}{2y^3}\right)^3 \div \frac{6x}{y^2}$

g $\left(\frac{6a^4}{b^6}\right)^2 \div \left(\frac{3a}{b^2}\right)^4$

h $\left(\frac{9a^4}{2b^2}\right)^2 \div \left(\frac{3b^2}{4a^3}\right)^3$

i $\left(\frac{3a^5}{4b^7}\right)^2 \div 4\left(\frac{1}{a^4b^4}\right)^7$

j $\left(\frac{3m^4}{n^3}\right)^2 \div \left(\frac{n^7}{2m^2}\right)^3$

k $\frac{1}{2}\left(\frac{p^2q}{q^5}\right)^3 \times \left(\frac{p}{q^8}\right)^3$

l $\frac{7}{8}\left(\frac{2r}{3r^3}\right)^3 \div \left(\frac{r^4}{s^2}\right)^3$

3 Simplify each of the following.

a $\left((x^3)^4\right)^2$

b $\left((2x^3)^2\right)^2$

c $\left((x^2y)^6\right)^4$

d $\left(\left(\frac{a^3}{b^2}\right)^2\right)^4$

e $\left(\left(\frac{a^3}{b^4c}\right)^2\right)^3$

f $\left(\left(\frac{a^9}{3b^7}\right)^2\right)^2$

Challenge Yourself

1 Simplify each of the following.

a $\left(\frac{4a^3b^2c^4}{2a^2bc}\right)^2 \times \left(\frac{3a^2b}{c^4}\right)^2$

b $\frac{(4a^2b^2)^2 \times -2(a^4b)^2}{(-3a^3b^2)^2}$

2 Write the following in simplest index notation.

a $\frac{(3x^4)^3}{(2y^2z^3)^2} \times \left(\frac{2xz}{x^3z}\right)^2$

b $\frac{(-5x^3y^2)^3}{(z^2)^2} \times \left(\frac{2x^3z}{5y^3}\right)^2$

3 Simplify the following.

a $\left(\frac{-9m^3n}{2mp^3}\right)^2 \div \left(\frac{-3mn^3}{p^2}\right)^2$

b $\frac{6}{7}\left(\frac{7r^4st^4}{2s^4t^3}\right)^2 \div \left(\frac{-6rs}{r^2}\right)^2$

Negative Powers: Round Three (Flipping Things Around)

Now we're up to laws 7 and 8.

These ones are about negative powers.

Here are the laws.

Exponent Laws

Law 7 Negative Power:
Take the reciprocal,
raise to positive power

$$a^{-n} = \frac{1}{a^n}$$

or

$$\frac{1}{a^{-n}} = a^n$$

Law 8 Negative Power of a Fraction:
Flip fraction, raise to positive power

$$\left(\frac{a}{b}\right)^{-n} = \left(\frac{b}{a}\right)^n$$

Let's go over these so everything makes sense.

First off.

What's a reciprocal?

A reciprocal is when you rewrite something as '1 over that thing'.

For a fraction, you just flip the top and bottom.

Second.

As soon as you see anything raised to a negative power.

Rewrite it with a positive power.

If it's a letter or a number, put 1 over it and change the power to positive.

If it's a fraction, flip it and change the power to positive.

Now, as you get to the harder stuff.

You're gonna see positive powers.

Negative powers.

Fractions.

Brackets with powers.

All of it.

In one expression.

And if you're not careful, you'll mess it up.

Before you start simplifying.

What you want to do is make all the powers positive.

It makes things a whole lot easier.

So, here's a trick to easily change negative powers to positive.

Works every time.

You don't even have to think about it.

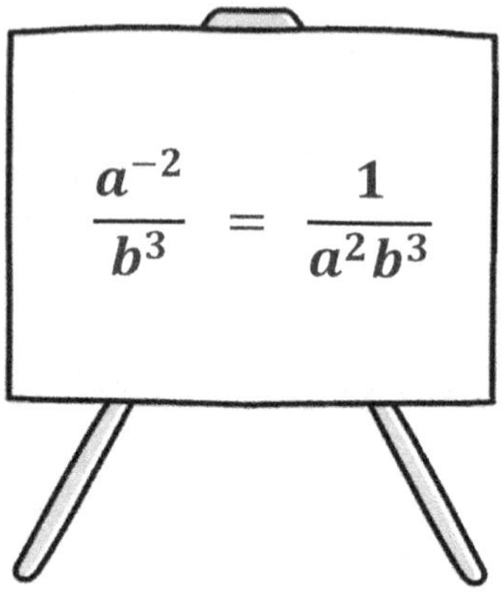

Change the Level.

Change the Sign.

This means you can move anything with a power from the top to the bottom of a fraction.

Or from the bottom to the top.

And when you do, just change the sign of the power.

Once everything's in positive powers, that's when you start simplifying.

Using the laws you've learned so far.

Got it?

Let's see how this plays out with some examples.

Worked Example 1

Evaluate each of the following.

a 4^{-3} b 7^{-2} c $\left(\frac{5}{2}\right)^{-3}$

Explanation	**Solution**
a Write the expression.	4^{-3}
Apply Law 7. (Take reciprocal, raise to positive power)	$=\frac{1}{4^3}$
Evaluate.	$=\frac{1}{64}$
b Write the expression.	7^{-2}
Apply Law 7.	$=\frac{1}{7^2}$
Evaluate.	$=\frac{1}{49}$
c Write the expression.	$\left(\frac{5}{2}\right)^{-3}$
Apply Law 8. (Flip fraction, raise to positive power)	$=\left(\frac{2}{5}\right)^{3}$
Apply Law 6 and evaluate. (Raise numerator and denominator to power separately)	$=\frac{8}{125}$

Worked Example 2

Evaluate each of the following.

a $\frac{1}{2^{-3}}$ b $\frac{1}{3^{-2}}$ c $\frac{6}{5^{-1}}$

Explanation	**Solution**
a Write the expression.	$\frac{1}{2^{-3}}$
Apply Law 7.	$= \frac{2^3}{1}$
Evaluate.	$= 8$
b Write the expression.	$\frac{1}{3^{-2}}$
Apply Law 7.	$= \frac{3^2}{1}$
Evaluate.	$= 9$
c Write the expression.	$\frac{6}{5^{-1}}$
Apply Law 7.	$= 6 \times 5^1$
Evaluate.	$= 30$

Worked Example 3

Express the following using positive powers.

a a^{-6} b $2a^{-4}$ c $\frac{4}{5}a^{-2}$ d $\frac{1}{a^{-9}}$

Explanation	**Solution**
a Write the expression.	a^{-6}
Apply Law 7.	$= \frac{1}{a^6}$
b Write the expression.	$2a^{-4}$
Apply Law 7.	$= \frac{2}{a^4}$
c Write the expression.	$\frac{4}{5}a^{-2}$
Apply Law 7.	$= \frac{4}{5} \times \frac{1}{a^2}$
Simplify.	$= \frac{4}{5a^2}$
d Write the expression.	$\frac{1}{a^{-9}}$
Apply Law 7.	$= \frac{a^9}{1}$
	$= a^9$

Worked Example 4

Express the following using positive powers.

a $3x^2y^{-5}$ b $x^3y^{-2} \times x^{-5}y^{-1}$ c $3(x^7y^{-4})^{-2}$

Explanation		Solution
a	Write the expression.	$3x^2y^{-5}$
	Apply Law 7 to term with negative power.	$= 3x^2 \times \frac{1}{y^5}$
	Multiply across and simplify.	$= \frac{3x^2}{y^5}$
b	Write the expression.	$x^3y^{-2} \times x^{-5}y^{-1}$
	Apply Law 7 to each term.	$= \frac{x^3}{y^2} \times \frac{1}{x^5y}$
	Simplify by applying Law 3.	$= \frac{1}{y^2} \times \frac{1}{x^2y}$
	Multiply across and apply Law 2.	$= \frac{1}{x^2y^3}$
c	Write the expression.	$3(x^7y^{-4})^{-2}$
	Apply Law 5 to bracket term. (Raise each term to power separately)	$= 3x^{-14}\,y^8$
	Apply Law 7 to term with negative power.	$= \frac{3y^8}{x^{14}}$

Worked Example 5

Express the following using positive powers.

a $a^{-2}b^4 \times (a^3b^{-4})^{-1}$ b $(2x^2y^3)^2 \times x^{-1}y^{-3}$

Explanation	Solution
a Write the expression.	$a^{-2}b^4 \times (a^3b^{-4})^{-1}$
Apply Law 7 to first term and Law 5 to bracket term.	$= \frac{b^4}{a^2} \times a^{-3}b^4$
Apply Law 7 to second term.	$= \frac{b^4}{a^2} \times \frac{b^4}{a^3}$
Apply Law 2.	$= \frac{b^{4+4}}{a^{2+3}}$
Simplify.	$= \frac{b^8}{a^5}$
b Write the expression.	$(2x^2y^3)^2 \times x^{-1}y^{-3}$
Apply Law 5 to bracket term and Law 7 on second term.	$= 2^2\, x^4y^6 \times \frac{1}{xy^3}$
Multiply.	$= \frac{4x^4y^6}{xy^3}$
Apply Law 3 to simplify.	$= 4x^3y^3$

Worked Example 6

Express the following using positive powers.

a $\left(\dfrac{ab^3}{c^2}\right)^{-4}$

b $\left(\dfrac{4a^3}{3b^{-2}}\right)^{-3}$

Explanation	Solution
a Write the expression.	$\left(\dfrac{ab^3}{c^2}\right)^{-4}$
Apply Law 8. (Flip fraction, raise to positive power)	$= \left(\dfrac{c^2}{ab^3}\right)^{4}$
Apply Laws 5 and 6.	$= \dfrac{c^8}{a^4b^{12}}$
b Write the expression.	$\left(\dfrac{4a^3}{3b^{-2}}\right)^{-3}$
Inside brackets, move negative power term to the top.	$= \left(\dfrac{4a^3b^2}{3}\right)^{-3}$
Apply Law 7.	$= \left(\dfrac{3}{4a^3b^2}\right)^{3}$
Apply Laws 5 and 6 and simplify.	$= \dfrac{3^3}{4^3a^9b^6}$
Evaluate numbers.	$= \dfrac{27}{64a^9b^6}$

Worked Example 7

Express the following using positive powers.

a $2^{-2}p^{-3}q^{-4}$ b $\dfrac{5^3p^3}{p^{-1}q^{-2}}$

Explanation		Solution
a	Write the expression.	$2^{-2}p^{-3}q^{-4}$
	Apply Law 7 to each term.	$=\dfrac{1}{2^2}\times\dfrac{1}{p^3}\times\dfrac{1}{q^4}$
	Multiply fractions and simplify.	$=\dfrac{1}{4p^3q^4}$
b	Write the expression.	$\dfrac{5^3p^3}{p^{-1}q^{-2}}$
	Apply Law 7 to terms with negative power and move to the top.	$=\dfrac{125\times p^3\times p\times q^2}{1}$
	Simplify.	$=125p^4q^2$

Forgetting to change the sign of the power when moving a term from the numerator to the denominator, or vice versa. ⚠

Worked Example 8

Simplify and write your answer with positive powers.

a $\dfrac{(-2)^3 \times 2^{-4}}{2^{-3}}$

b $\dfrac{\left(x^{-2}\right)^3 \times \left(y^4\right)^{-2}}{x^{-5} \times (y^{-2})^3}$

Explanation	Solution
a Write the expression.	$\dfrac{(-2)^3 \times 2^{-4}}{2^{-3}}$
Apply Law 5 to backet term.	$= \dfrac{-8 \times 2^{-4}}{2^{-3}}$
Apply Law 7.	$= \dfrac{-8 \times 2^{3}}{2^{4}}$
Simplify and evaluate.	$= \dfrac{-8 \times 8}{16}$
	$= -4$
b Write the expression.	$\dfrac{\left(x^{-2}\right)^3 \times (y^4)^{-2}}{x^{-5} \times (y^{-2})^3}$
Apply Law 5 to backet terms.	$= \dfrac{x^{-6} \times y^{-8}}{x^{-5} \times y^{-6}}$
Apply Law 7.	$= \dfrac{x^{5} \times y^{6}}{x^{6} \times y^{8}}$
Apply Law 3 and simplify.	$= \dfrac{1}{xy^2}$

Worked Example 9

Simplify $\dfrac{8x^3y^{-4}}{(6xy^2)^3} \div \dfrac{4x^{-2}y^{-4}}{6x^{-5}y}$

Explanation	Solution
Write the expression	$\dfrac{8x^3y^{-4}}{(6xy^2)^3} \div \dfrac{4x^{-2}y^{-4}}{6x^{-5}y}$
Apply Law 5 to bracket term.	$= \dfrac{8x^3y^{-4}}{6^3x^3y^6} \div \dfrac{4x^{-2}y^{-4}}{6x^{-5}y}$
Apply Law 7 to rewrite all terms with positive powers.	$= \dfrac{8x^3}{6^3x^3y^6y^4} \div \dfrac{4x^5}{6yx^2y^4}$
Apply Law 2 to simplify each denominator.	$= \dfrac{8x^3}{6^3x^3y^{10}} \div \dfrac{4x^5}{6x^3y^5}$
Apply Law 3 to simplify both fraction.	$= \dfrac{8}{6^3y^{10}} \div \dfrac{4x^3}{6y^5}$
Flip the second fraction. Change the division sign to multiplication.	$= \dfrac{8}{6^3y^{10}} \times \dfrac{6y^5}{4x^3}$
Apply Law 3 and cancel out common factors.	$= \dfrac{2}{36y^5} \times \dfrac{1}{x^3}$
Multiply fractions and simplify.	$= \dfrac{1}{18x^3y^5}$

If a number is multiplied to a letter with a negative power, the power only applies to the letter. Unless everything's in brackets.

Your Turn To Try

1 $\frac{1}{8}$ **is the same as?**

A 2^3

B 2^{-3}

C 4^2

D 4^{-2}

2 $\frac{1}{x^{-4}}$ **is the same as?**

A $4x$

B $-4x$

C x^4

D $-x^4$

3 **When simplified** $\left(2a^6b^{-4}\right)^{-3}$ **is equal to?**

A $\frac{2b^{12}}{a^{18}}$

B $\frac{8b^{12}}{a^{18}}$

C $\frac{b^{12}}{8a^{18}}$

D $\frac{b^{12}}{6a^{18}}$

4 **Expressed in positive powers this** $\left(\frac{3^4}{2^3}\right)^{-2}$ **is equal to?**

A $\frac{3^8}{2^6}$

B $\frac{2^6}{3^8}$

C $\frac{1}{3^8 2^6}$

D $\frac{2^9}{3^6}$

5 $\left(\frac{x^2y}{y^{-3}}\right)^2$ **is equal to?**

A x^4y^8

B x^4y^6

C x^2y^8

D x^2y^6

Question	1	2	3	4	5
Answer	B	C	C	B	A

Time To Practice

1 Simplify these expressions with positive exponents. For numbers with powers calculate its value.

a 4^{-1}

b 3^{-3}

c 5^{-3}

d $2^{-3} \times 2^2$

e $7^{-4} \times 7^2$

f $\dfrac{3}{4^{-2}}$

g $\dfrac{1}{a^{-3}}$

h $b^{-4} \times b$

i $c^{-5} \div c^{-3}$

j $\dfrac{1}{y^{-4} \times y}$

k $x^{-3} \times y \times x^2$

l $x^{-4} \times y^{-3} \times z^{-2}$

2 Evaluate or express each of the following using positive powers.

a $\left(\dfrac{5}{4}\right)^{-3}$

b $\left(\dfrac{2}{3}\right)^{-2}$

c $\left(\dfrac{1}{2^{-3}}\right)^{-2}$

d $\left(\dfrac{x^2}{y^{-3}}\right)^{4}$

e $\left(\dfrac{2x^{-3}}{y^{-2}}\right)^{-1}$

f $\left(\dfrac{2}{x}\right)^{-3} \times x^{-2}$

g $\dfrac{p^{-3}q^2}{r^{-4}p^2}$

h $\dfrac{(p^{-1})^3 \times (q^3)^{-3}}{p^{-6} \times (q^{-5})^3}$

i $\dfrac{(p^2)^{-2} \times (q^{-3})^{-2}}{p^4 \times (q^{-1})^2}$

j $\dfrac{x^{-3}y^{-4}}{z^{-2}} \times \dfrac{x^{-2}y^3}{z^{-2}}$

k $\left(\dfrac{x^{-2}y}{z^2}\right)^3 \times \left(\dfrac{z}{x^{-1}}\right)^{-3}$

l $\left(\dfrac{rs}{r^{-5}}\right)^{-3} \div \left(\dfrac{2r^{-8}}{s^2}\right)^{-2}$

3 Simplify and express each of the following using positive powers.

a $\left(\dfrac{x^{-3}}{2^{-2}y^{-4}}\right)^{-2}$

b $\left(\dfrac{m^{-2}n^{-3}}{3p^2}\right)^{-2} \div p^{-6}$

c $\left(\dfrac{2^{-2}x^{-1}}{y^3}\right)^{-2} \div y^{-9}$

Challenge Yourself

1 Simplify each of the following. Express your answer using positive powers.

a $\dfrac{2^{-4}x^{-3}y^2}{z^{-2}} \times \dfrac{x^6y^4}{z^{-4}}$

b $\dfrac{2^{-3}x^2y^{-4}}{z^{-3}} \times \left(\dfrac{x^{-1}z}{y^{-2}}\right)^{-2}$

2 Simplify each of the following. Express your answer using positive powers.

a $\dfrac{(m^{-4})^3 \times (n^{-3})^2}{(2m^2n^3)^2} \times \left(\dfrac{mn}{3m^{-3}}\right)^{-2}$

b $\dfrac{4}{7}\left(\dfrac{m^3n^{-4}}{p^{-2}}\right)^{-2} \times \left(\dfrac{p^2m^{-1}}{n^3}\right)^{-3}$

3 Simplify each of the following. Express your answer using positive powers.

a $\dfrac{2}{5}\left(\dfrac{3x^{-3}y^{-1}}{x^{-2}y^{-4}z}\right)^{-2} \div \left(\dfrac{x^2y^{-1}}{4y^5z^3}\right)^{-1}$

b $\left(\dfrac{x^{-2}y^{-1}}{x^{-4}y^{-1}z^{-5}}\right)^{-2} \div \left(\dfrac{x^{-2}y^{-5}z^{-9}}{3^{-1}y^{-8}z^{-1}}\right)^{-1}$

Fractional Powers: Final Round (Don't Forget Your Roots)

Last exponent law.

This one's for fractional powers.

Law 9.

Here it is.

Exponent Laws

Law 9 Fractional Power: Take the root, then raise to the power

$$a^{\frac{m}{n}} = a^{\frac{1}{n} \times m} = \left(a^{\frac{1}{n}}\right)^m$$

or

$$a^{\frac{m}{n}} = \sqrt[n]{a^m}$$

OK, this looks a bit crazy.

You've got letters, fractions, brackets, and roots all over the place.

But there's a way to make sense of it.

Look at the first rule.

The fractional power gets split into two parts that multiply together.

A fractional part, which is 1 over its denominator.

Times by a whole number, which is the numerator.

From there, applying Law 4 backwards.

The base gets raised to the fractional part inside brackets.

Then those brackets get raised to the whole number.

Now, if the original power already has a numerator of 1.

No need to split it.

We only split it to get a fractional power with a numerator of 1.

Because that's when we can work it out.

So, how do we work it out exactly?

We'll go over the steps after we look at the second rule.

It all ties together.

Look at the second rule.

$$a^{\frac{m}{n}} = \sqrt[n]{a^m}$$

The same fractional power is written in root form.

The base number goes under the root sign.

The numerator becomes the power of the base number.

The denominator goes to the left, outside the root sign.

Now, what's a root?

A root is a number that, when multiplied by itself, a certain number of times, equals what's inside the root sign.

Okay, here's how you workout a number raised to a fractional power, using the first rule.

Make sure the numerator is 1.

Then look at the denominator.

That number tells you how many times you multiply something by itself to get the base number.

If the numerator isn't 1.

Split the power and write it in bracket form.

Now you can work out what's inside the bracket, because the numerator is 1.

Once you've done that.

Raise it to the power outside the brackets.

As for using the second rule, when it's in root form.

It's pretty much the same thing.

Look at the number outside the root sign.

That tells you how many times to multiply something by itself to get the number inside.

With fractional powers, you can use either rule.

They're just two ways to express the same thing.

So, which one do you use?

Generally.

If the base is a letter, use fractional powers.

If the base is a number, the root form's more common.

But honestly, you need to be able to switch between the two.

Going from fractional form to the root form.

Or vice versa.

Now, this skill.

Being able to switch between the two forms.

Really, *really* important.

A lot of people can't do it, and because of that, they get stuck.

Not just in algebra, but in other areas of math too.

So practice and get good at switching between the two forms.

Are you clear on what to do with fractional powers?

You need to find the nth root of a number., where $\boldsymbol{n}$ is the denominator of the power.

How do you do that?

Easy, use this table.

Power Table

Root	2nd Power	3rd Power	4th Power	5th Power
x	x^2	x^3	x^4	x^5
1	1	1	1	1
2	4	8	16	32
3	9	27	81	243
4	16	64	256	1024
5	25	125	625	3,125
6	36	216	1,296	7,776
7	49	343	2,401	16,807
8	64	512	4,096	32,768
9	81	729	6,561	59,049
10	100	1,000	10,000	100,000

Let's say you need the 4th root of 81.

You're looking for a number that, when multiplied by itself 4 times, equals 81.

So, here's what you do.

Look at the 4th Power column and find 81.

Then look at its root number.

It's 3.

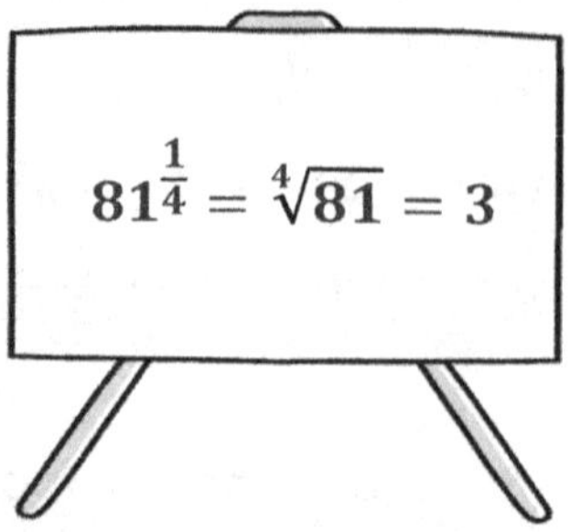

That means the 4th root of 81 is 3.

In other words, 3 multiplied by itself 4 times equals 81.

Most of the numbers and roots you'll need are in this table.

Especially numbers less than 1,024.

You'll see them a lot.

Eventually, you'll recognize them and know their roots.

As for numbers bigger than 1,024?

You'll probably be allowed to use a calculator to find their roots.

Alright, a couple of little things to note.

The 2nd root, when the power is $\frac{1}{2}$, is called the *square root*.

And the 3rd root, when the power is $\frac{1}{3}$, is called the *cube root*.

We usually don't say 2nd or 3rd root.

We say square root or cube root.

The nth root means you multiply something by itself n times to get the base number.

This n is equal to the denominator of the fractional power.

Also.

When the power is $\frac{1}{2}$ and you switch to root form.

Don't write a small 2 to the left of the root sign.

Just leave it blank.

But for any other fractional power, write the denominator to the left of the root sign.

OK.

Now for the fun part.

Let's try out some examples.

Worked Example 1

Express each of the following in root form.

a $3^{\frac{1}{5}}$ b $6^{\frac{1}{3}}$ c $5^{\frac{2}{3}}$ d $11^{\frac{2}{7}}$

Explanation	Solution
a Write the expression.	$3^{\frac{1}{5}}$
Apply Law 9. (Take the root, then raise to the power)	$= \sqrt[5]{3}$
b Write the expression.	$6^{\frac{1}{3}}$
Apply Law 9.	$= \sqrt[3]{6}$
c Write the expression.	$5^{\frac{2}{3}}$
Apply Law 9.	$= \sqrt[3]{5^2}$
Simplify.	$= \sqrt[3]{25}$
d Write the expression.	$11^{\frac{2}{7}}$
Apply Law 9.	$= \sqrt[7]{11^2}$
Simplify.	$= \sqrt[7]{121}$

The small number in front of the root sign, called the *index*, is the denominator in the fractional power. It tells you which n^{th} root to find.

Worked Example 2

Evaluate the following:

a $8^{\frac{2}{3}}$ b $16^{\frac{3}{4}}$ c $36^{\frac{3}{2}}$

Explanation	Solution
a Write the expression.	$8^{\frac{2}{3}}$
Apply Law 9.	$=\left(8^{\frac{1}{3}}\right)^2$
Find the cube root, then square it.	$=(2)^2$
Simplify.	$=4$
b Write the expression.	$16^{\frac{3}{4}}$
Apply Law 9.	$=\left(16^{\frac{1}{4}}\right)^3$
Find the 4th root, then cube it.	$=(2)^3$
Simplify.	$=8$
c Write the expression.	$36^{\frac{3}{2}}$
Apply Law 9.	$=\left(36^{\frac{1}{2}}\right)^3$
Find the square root, then cube it.	$=(6)^3$
Simplify.	$=216$

Worked Example 3

Rewrite the following in fractional power form without evaluating.

a $\sqrt{3}$ b $\sqrt[3]{13}$ c $\sqrt[4]{x^3}$ d $\sqrt{5x^4}$

Explanation	**Solution**
a Write the expression.	$\sqrt{3}$
Apply Law 9.	$= 3^{\frac{1}{2}}$
b Write the expression.	$\sqrt[3]{13}$
Apply Law 9.	$= 13^{\frac{1}{3}}$
c Write the expression.	$\sqrt[4]{x^3}$
Apply Law 9.	$= (x^3)^{\frac{1}{4}}$
Apply Law 4. (Keep base, multiply powers)	$= x^{\frac{3}{4}}$
d Write the expression.	$\sqrt{5x^4}$
Apply Law 9.	$= (5x^4)^{\frac{1}{2}}$
Apply Law 5. (Raise each term to power separately)	$= 5^{\frac{1}{2}}x^2$

Worked Example 4

Evaluate the following.

a $\sqrt{100}$ b $\sqrt[3]{64}$ c $\sqrt[4]{81}$ d $\sqrt[5]{32}$

Explanation	Solution
a Write the expression.	$\sqrt{100}$
Simplify.	$= 10$
b Write the expression.	$\sqrt[3]{64}$
Apply Law 9.	$= 64^{\frac{1}{3}}$
Simplify.	$= 4$
c Write the expression.	$\sqrt[4]{81}$
Apply Law 9.	$= 81^{\frac{1}{4}}$
Simplify.	$= 3$
d Write the expression.	$\sqrt[5]{32}$
Apply Law 9.	$= 32^{\frac{1}{5}}$
Simplify.	$= 2$

If there's just one number under the root, you don't need to rewrite it as a fractional power. Simplify it in one step.

Worked Example 5

Use exponent laws to simplify the following:

a $\sqrt{25a^4}$ b $\sqrt[3]{27b^6}$

Explanation	Solution
a Write the expression.	$\sqrt{25a^4}$
Apply law 9.	$= (25a^4)^{\frac{1}{2}}$
Apply law 5.	$= 25^{\frac{1}{2}} \times a^{\frac{4}{2}}$
Simplify and combine terms.	$= 5 \times a^2$
	$= 5a^2$
b Write the expression.	$\sqrt[3]{27b^6}$
Apply law 9.	$= (27b^6)^{\frac{1}{3}}$
Apply law 5.	$= 27^{\frac{1}{3}} \times b^{\frac{6}{3}}$
Simplify and combine terms.	$= 3 \times b^2$
	$= 3b^2$

Not finding the root of each number and letter that's multiplied under the root sign. ⚠

Worked Example 6

Express each of the following using fractional power form.

a $\dfrac{1}{\sqrt{x^5}}$ b $\dfrac{1}{\sqrt[5]{y}}$ c $\dfrac{1}{\sqrt{z^{-3}}}$

Explanation	Solution
a Write the expression.	$\dfrac{1}{\sqrt{x^5}}$
Apply Law 9 to denominator.	$= \dfrac{1}{(x^5)^{\frac{1}{2}}} = \dfrac{1}{x^{\frac{5}{2}}}$
Apply Law 7. (Take reciprocal, raise to positive power)	$= x^{-\frac{5}{2}}$
b Write the expression.	$\dfrac{1}{\sqrt[5]{y}}$
Apply Law 9 to denominator.	$= \dfrac{1}{y^{\frac{1}{5}}}$
Apply Law 7.	$= y^{-\frac{1}{5}}$
c Write the expression.	$\dfrac{1}{\sqrt{z^{-3}}}$
Apply Law 9 and Law 4 to denominator.	$= \dfrac{1}{(z^{-3})^{\frac{1}{2}}} = \dfrac{1}{z^{-\frac{3}{2}}}$
Apply Law 7.	$= z^{\frac{3}{2}}$

Worked Example 7

Simplify each of the following, expressing your answer with positive powers.

a $\dfrac{1}{\sqrt{x^{-4}}}$

b $\dfrac{1}{\sqrt[3]{8x^9}}$

Explanation	Solution
a Write the expression.	$\dfrac{1}{\sqrt{x^{-4}}}$
Apply Law 9 and Law 4 to denominator.	$=\dfrac{1}{(x^{-4})^{\frac{1}{2}}}=\dfrac{1}{x^{-2}}$
Apply Law 7.	$=x^2$
b Write the expression.	$\dfrac{1}{\sqrt[3]{8x^9}}$
Apply Law 9 to denominator.	$=\dfrac{1}{(8x^9)^{\frac{1}{3}}}$
Apply Law 5.	$=\dfrac{1}{8^{\frac{1}{3}}x^3}$
Simplify.	$=\dfrac{1}{2x^3}$

For negative fractional powers, apply Law 7 to change the power to positive before simplifying or evaluating.

Worked Example 8

Evaluate the following without a calculator:

a $\mathbf{9^{-\frac{1}{2}}}$ b $\mathbf{32^{-\frac{1}{5}}}$ c $\mathbf{81^{-\frac{1}{4}}}$

Explanation	**Solution**
a Write the expression.	$9^{-\frac{1}{2}}$
Apply Law 7.	$= \frac{1}{9^{\frac{1}{2}}}$
Apply Law 9.	$= \frac{1}{3}$
b Write the expression.	$32^{-\frac{1}{5}}$
Apply Law 7.	$= \frac{1}{32^{\frac{1}{5}}}$
Apply Law 9.	$= \frac{1}{2}$
c Write the expression.	$81^{-\frac{1}{4}}$
Apply Law 7.	$= \frac{1}{81^{\frac{1}{4}}}$
Apply Law 9.	$= \frac{1}{3}$

Worked Example 9

Use exponent laws to simplify the following:

a $\left(343a^9b^6\right)^{\frac{1}{3}}$

b $\left(\frac{32}{a^{10}}\right)^{-\frac{2}{5}}$

Explanation	Solution
a Write the expression.	$\left(343a^9b^6\right)^{\frac{1}{3}}$
Apply Law 5.	$= 343^{\frac{1}{3}} \times a^{\frac{9}{3}} \times b^{\frac{6}{3}}$
Apply Law 9.	$= 7 \times a^3 \times b^2$
	$= 7a^3b^2$
b Write the expression.	$\left(\frac{32}{a^{10}}\right)^{-\frac{2}{5}}$
Apply Law 8. (Flip fraction, raise to positive power)	$= \left(\frac{a^{10}}{32}\right)^{\frac{2}{5}}$
Apply Law 6.	$= \frac{a^{\frac{20}{5}}}{32^{\frac{2}{5}}} = \frac{a^4}{\left(32^{\frac{1}{5}}\right)^2}$
Simplify the denominator.	$= \frac{a^4}{2^2}$
	$= \frac{a^4}{4}$

Forgetting which part of the fraction becomes the root (denominator) and which becomes the power (numerator). ⚠

Worked Example 10

Simplify $\left(x^{-2}y^{\frac{1}{2}}\right)^2 \times \left(2x^{-\frac{1}{5}}y^{-\frac{1}{2}}\right)^5$

Explanation	**Solution**
Write the expression.	$\left(x^{-2}y^{\frac{1}{2}}\right)^2 \times \left(2x^{-\frac{1}{5}}y^{-\frac{1}{2}}\right)^5$
Apply Law 5 to each bracket.	$= x^{-4}y \times 2^5x^{-1}y^{-\frac{5}{2}}$
Apply Law 7. (Keep base, add powers)	$= \frac{y}{x^4} \times \frac{32}{xy^{\frac{5}{2}}}$
Apply law 2. Multiply across.	$= \frac{32y}{x^5y^{\frac{5}{2}}}$
Apply Law 3 to simplify.	$= \frac{32}{x^5y^{\frac{3}{2}}}$

It's easy to miss details with harder expressions.
So, write down each step to avoid mistakes.

Your Turn To Try

1 **What is the value of $16^{\frac{1}{4}}$?**

A 2 B 4

C 8 D 16

2 **Which of the following gives the correct value of $32^{\frac{3}{5}}$?**

A 32 B 16

C 8 D 4

3 **What is the simplified form of $\left(\frac{64}{x^3}\right)^{\frac{1}{3}}$?**

A $\frac{x}{4}$ B $\frac{4}{x}$

C $\frac{x^2}{4}$ D $\frac{4}{x^2}$

4 **Which expression is equal to $(81y^{16})^{\frac{1}{4}}$?**

A $9y^4$ B $3y^6$

C $9y^6$ D $3y^4$

5 **When simplified $(a^9b^6)^{\frac{2}{3}}$ is ?**

A a^3b^2 B a^6b^4

C a^4b^3 D a^4b^6

Question	1	2	3	4	5
Answer	A	C	B	D	B

Time To Practice

1 Express each of the following in root form, then evaluate.

a $27^{\frac{1}{3}}$ b $121^{\frac{1}{2}}$ c $49^{\frac{1}{2}}$

d $8^{\frac{5}{3}}$ e $25^{\frac{3}{2}}$ f $49^{\frac{3}{2}}$

g $16^{\frac{3}{4}}$ h $4^{-\frac{1}{2}}$ i $27^{-\frac{2}{3}}$

j $8^{-\frac{1}{3}}$ k $64^{-\frac{2}{3}}$ l $(125^{-1})^{\frac{1}{3}}$

2 Evaluate each of the following.

a $\left(\frac{9}{16}\right)^{-\frac{1}{2}}$ b $\left(\frac{16}{81}\right)^{-\frac{3}{4}}$ c $\left(\frac{27}{64}\right)^{-\frac{2}{3}}$

3 Express the following in exponential form. (Do not evaluate)

a $\sqrt{x^3}$ b $\sqrt[3]{x^5}$ c $\sqrt[4]{16x}$

d $\sqrt[3]{y^2}$ e $\sqrt[6]{64y^7}$ f $\sqrt[3]{8y^4}$

4 Simplify and express each of the following using positive powers.

a $(x^6)^{\frac{1}{3}}$ b $(25x^2)^{\frac{1}{2}}$ c $(16x^8)^{\frac{1}{4}}$

d $(a^2b^4)^{\frac{1}{2}}$ e $(8a^3b^3)^{\frac{1}{3}}$ f $3a^{\frac{1}{2}} \times 2a^{-\frac{3}{4}}$

g $(a^8)^{-\frac{1}{2}}$ h $(8x^9)^{-\frac{1}{3}}$ i $(x^4y^8)^{-\frac{3}{4}}$

j $(x^2)^{\frac{1}{2}} \times x^{-\frac{5}{3}}$ k $\left(\frac{x^{12}y^{-8}}{z^4}\right)^{\frac{3}{4}}$ l $\left(\frac{x^6}{y^{-9}}\right)^{\frac{2}{3}}$

Challenge Yourself

1 Simplify each of the following. Express your answer using positive powers.

a $\left(a^2b^4\right)^{-\frac{1}{2}} \times \left(ba^2\right)^{\frac{3}{2}}$

b $\left(a^3b^{-2}\right)^{-\frac{2}{3}} \times \left(a^{-1}b^2\right)^{\frac{1}{3}}$

2 Simplify each of the following. Express your answer using positive powers.

a $\left(2^{-1}x^{\frac{3}{2}}y^{\frac{5}{4}}\right)^2 \times \left(8x^{-\frac{7}{4}}y^{\frac{1}{2}}\right)^{\frac{2}{3}}$

b $\left(x^{\frac{2}{5}}y^{-\frac{1}{4}}\right)^2 \div \left(x^{-\frac{3}{2}}y^{\frac{1}{3}}\right)^{\frac{2}{3}}$

3 Simplify each of the following. Express your answer using positive powers.

a $\left(\dfrac{a^{\frac{1}{2}}b^{\frac{1}{3}}}{2c^{-\frac{1}{2}}}\right)^2 \times \left(a^{-1}c^{\frac{1}{3}}\right)^{-1}$

b $\left(\dfrac{2a^2b^{-1}}{c^{\frac{1}{2}}}\right)^{-2} \div \left(\dfrac{4a^{-1}b^2}{c}\right)^{\frac{1}{2}}$

2. Answers

Multiplication and Division of Exponents

Time to Practice

1	a	32	b	27	c	125
	d	x^6	e	y^5	f	z^{16}
	g	a^{13}	h	b^{10}	i	c^{19}
	j	5	k	7	l	-1
	m	$12a^6$	n	$20b^9$	o	$35c^7$
2	a	49	b	64	c	1000
	d	m^7	e	n^{15}	f	p^{12}
	g	25	h	81	i	64
	j	$15x^6$	k	$2y^4$	l	$10z^{10}$
	m	$6a^5$	n	$9b^{11}$	o	$4c^8$

Challenge Yourself

1	a	$2u^7v^7$	b	$4e^9f^2$
2	a	$-3a^4b^3$	b	$9ab^2$
3	a	$\frac{7x^7y^2}{4}$	b	$\frac{16y^3z^4}{27}$

Power of a Power

Time to Practice

1. a a^{10} b b^{16} c $8a^{18}$
   d $-27x^{12}$ e $x^{12}y^5$ f x^4y^{11}
   g $4x^5y^{10}$ h $16x^{18}y^{11}$ i y^{12}
   j $\frac{m^{13}n}{2}$ k $\frac{32pq^3}{9}$ l $54r^9s^7$

2. a $\frac{a^2}{b^4}$ b $\frac{a^{10}}{b^{15}}$ c $\frac{16a^6}{b^4}$
   d $\frac{x^{23}}{y^{17}}$ e $\frac{x^{18}}{4y^8}$ f $\frac{9x^5}{16y^7}$
   g $\frac{4a^4}{9b^4}$ h $\frac{48a^{17}}{b^{10}}$ i $\frac{9a^{38}b^{14}}{64}$
   j $\frac{72m^{14}}{n^{27}}$ k $\frac{p^9}{2q^{36}}$ l $\frac{7s^6}{27r^{18}}$

3. a x^{24} b $16x^{12}$ c $x^{48}y^{24}$
   d $\frac{a^{24}}{b^{16}}$ e $\frac{a^{18}}{b^{24}c^6}$ f $\frac{a^{36}}{81b^{28}}$

Challenge Yourself

1. a $\frac{36a^6b^4}{c^2}$ b $\frac{-32a^6b^2}{9}$

2. a $\frac{27x^8}{y^4z^6}$ b $\frac{-20x^{15}}{z^2}$

3. a $\frac{9m^2}{4p^2n^4}$ b $\frac{7r^{10}t^2}{24s^8}$

Negative Powers

Time to Practice

1	a	$\frac{1}{4}$	b	$\frac{1}{27}$	c	$\frac{1}{125}$
	d	$\frac{1}{2}$	e	$\frac{1}{49}$	f	48
	g	a^3	h	$\frac{1}{b^3}$	i	$\frac{1}{c^2}$
	j	y^3	k	$\frac{y}{x}$	l	$\frac{1}{x^4y^3z^2}$
2	a	$\frac{64}{125}$	b	$\frac{9}{4}$	c	$\frac{1}{64}$
	d	x^8y^{12}	e	$\frac{x^3}{2y^2}$	f	$\frac{x}{8}$
	g	$\frac{r^4q^2}{p^5}$	h	p^3q^6	i	$\frac{q^8}{p^8}$
	j	$\frac{z^4}{x^5y}$	k	$\frac{y^3}{x^9z^9}$	l	$\frac{4}{r^{34}s^7}$
3	a	$\frac{x^6}{16y^8}$	b	$9p^{10}m^4n^6$	c	$16x^2y^{15}$

Challenge Yourself

1	a	$\frac{x^3y^6z^6}{16}$	b	$\frac{x^4z}{8y^8}$
2	a	$\frac{9}{4m^{24}n^{14}}$	b	$\frac{4n^{17}}{7m^3p^{10}}$
3	a	$\frac{x^4}{90y^{12}z}$	b	$\frac{3y^3}{x^6z^{18}}$

Fractional Powers

Time to Practice

1	a	3	b	11	c	7
	d	32	e	125	f	343
	g	8	h	$\frac{1}{2}$	i	$\frac{1}{9}$
	j	$\frac{1}{2}$	k	$\frac{1}{16}$	l	$\frac{1}{5}$
2	a	$\frac{4}{3}$	b	$\frac{27}{8}$	c	$\frac{16}{9}$
3	a	$x^{\frac{3}{2}}$	b	$x^{\frac{5}{3}}$	c	$(16x)^{\frac{1}{4}}$
	d	$y^{\frac{2}{3}}$	e	$(64y^7)^{\frac{1}{6}}$	f	$(8y^4)^{\frac{1}{3}}$
4	a	x^2	b	$5x$	c	$2x^2$
	d	ab^2	e	$2ab$	f	$\frac{6}{a^{\frac{1}{4}}}$
	g	$\frac{1}{a^4}$	h	$\frac{1}{2x^3}$	i	$\frac{1}{x^3y^6}$
	j	$\frac{1}{x^{\frac{2}{3}}}$	k	$\frac{x^9}{y^6z^3}$	l	x^4y^6

Challenge Yourself

1	a	$\frac{a^2}{b^{\frac{1}{2}}}$	b	$\frac{b^2}{a^{\frac{7}{3}}}$
2	a	$x^{\frac{11}{6}}y^{\frac{17}{6}}$	b	$\frac{x^{\frac{9}{5}}}{y^{\frac{13}{8}}}$
3	a	$\frac{a^2b^{\frac{2}{3}}c^{\frac{2}{3}}}{4}$	b	$\frac{bc^{\frac{3}{2}}}{8a^{\frac{7}{2}}}$

Surds

Identifying Surds: Is It a Surd or Just Acting Like One?

OK, you understand fractional powers.

You know how to switch between them and root form.

And how to find the nth root of a number.

A surd is just a number in root form, or with a fractional power.

Except there's one big difference.

When you simplify it, you don't get a whole number, and you don't get a fraction either.

There's just no whole number that you can multiply by itself n times to get what's under the root.

If you check the power table, and look at a power column.

The number under the root sign won't be there.

That's when you know you've got a surd.

For instance.

Say you've got $\sqrt[4]{64}$.

Check the 4th Power column.

Do you see 64?

Nope.

That means there's no whole number, when multiplied by itself 4 times, equals 64.

Therefore $\sqrt[4]{64}$ is a surd.

Power Table

Root	2nd Power	3rd Power	4th Power	5th Power
x	x^2	x^3	x^4	x^5
2	4	8	16	32
3	9	27	81	243
4	16	64	256	1024
5	25	125	625	3,125
6	36	216	1,296	7,776
7	49	343	2,401	16,807
8	64	512	4,096	32,768
9	81	729	6,561	59,049
10	100	1,000	10,000	100,000

What about when there's a fraction under the root sign?

How would that work?

Because the power table only has whole numbers.

You can still use it.

The only difference is, you now look for *both* the numerator and the denominator.

In the same power column.

If they're both there.

The root is a fraction.

That's because the roots of the numerator and the denominator are whole numbers.

And when you have a whole number over a whole number.

You get a fraction.

Not a surd.

But if one or both aren't in the same power column.

Then what you've got is a surd.

Later, when we cover surd rules, you'll learn to split fractions under a root into two roots.

For now, just use the power table.

For example.

Let's say you've got $\sqrt[3]{\frac{8}{27}}$.

Check the 3rd Power column - do you see both 8 and 27 there?

Yes.

The cube roots of the top and bottom are 2 and 3.

This means $\frac{2}{3}$ when multiplied by itself 3 times, equals $\frac{8}{27}$.

So, $\sqrt[3]{\frac{8}{27}}$ is **not** a surd.

Worked Example 1

Determine which of the following numbers are surds.

a $\sqrt{16}$ b $\sqrt{13}$ c $\sqrt{\frac{1}{25}}$

Explanation	Solution
a Write the expression.	$\sqrt{16}$
Find a number that multiplies by itself to give the value under the square root.	$\sqrt{4 \times 4} = 4$
Is this number a fraction or a whole number?	4 is a whole number.
Conclusion.	$\therefore \sqrt{16}$ is NOT a surd.
b Write the expression.	$\sqrt{13}$
Find a number that multiplies by itself to give the value under the square root.	No number multiplied by itself equals 13.
Conclusion.	$\therefore \sqrt{13}$ is a surd.
c Write the expression.	$\sqrt{\frac{1}{25}}$
Find a fraction that multiplies by itself to give the value under the square root.	$= \sqrt{\frac{1}{5} \times \frac{1}{5}} = \frac{1}{5}$
Is this number a fraction or a whole number?	$\frac{1}{5}$ is a fraction.
Conclusion.	$\therefore \sqrt{\frac{1}{25}}$ is NOT a surd.

Worked Example 2

Determine which of the following numbers are surds.

a $\sqrt[3]{17}$ b $\sqrt[3]{27}$ c $\sqrt{0.01}$

Explanation		Solution
a	Write the expression.	$\sqrt[3]{17}$
	Find a number that multiplies by itself three times to give the value under the square root.	No number multiplied by itself three times equals 17.
	Conclusion.	$\therefore \sqrt[3]{17}$ is a surd.
b	Write the expression.	$\sqrt[3]{27}$
	Find a number that multiplies by itself three times to give the value under the square root.	$\sqrt[3]{27} = \sqrt[3]{3 \times 3 \times 3} = 3$
	Is this number a fraction or a whole number?	3 is a whole number.
	Conclusion.	$\therefore \sqrt[3]{27}$ is NOT a surd.
c	Write the expression.	$\sqrt{0.01}$
	Change the decimal to fraction form.	$= \sqrt{\frac{1}{100}}$
	Find a fraction that multiplies by itself to give the value under the square root.	$= \sqrt{\frac{1}{10} \times \frac{1}{10}} = \frac{1}{10}$
	Is this number a fraction or a whole number?	$\frac{1}{10}$ is a fraction.
	Conclusion.	$\therefore \sqrt{0.01}$ is NOT a surd.

Your Turn To Try

1 **Which of the following expressions is a surd?**

A $\sqrt{25}$ B $\sqrt{36}$

C $\sqrt{50}$ D $\sqrt{49}$

2 **Which of the following statements about the expression $\sqrt[3]{64}$ is NOT true?**

A It can be simplified B It's a surd

C It's a cube root D It's not a surd

3 **Given the expressions below, which one is NOT a surd?**

A $\sqrt[3]{27}$ B $\sqrt[3]{1000}$

C $\sqrt[3]{125}$ D $\sqrt[3]{20}$

4 **Identify the surd from the following expressions:**

A $\sqrt{0.001}$ B $\sqrt{0.04}$

C $\sqrt{0.09}$ D $\sqrt{0.81}$

5 **Which of the following expressions is a surd?**

A $\sqrt{\frac{4}{9}}$ B $\sqrt{\frac{9}{16}}$

C $\sqrt{\frac{16}{25}}$ D $\sqrt{\frac{5}{25}}$

Question	1	2	3	4	5
Answer	C	B	D	A	D

Time To Practice

1 Which of these are surds? Answer 'Yes' or 'No'.

a $\sqrt{10}$ b $\sqrt{144}$ c $\sqrt{55}$

d $\sqrt{18}$ e $\sqrt{8}$ f $\sqrt{1}$

g $\sqrt[3]{21}$ h $\sqrt[3]{216}$ i $\sqrt[3]{32}$

j $\sqrt[4]{16}$ k $\sqrt[4]{40}$ l $\sqrt[5]{32}$

m $\sqrt{\frac{1}{4}}$ n $\sqrt{\frac{4}{49}}$ o $\sqrt{\frac{16}{32}}$

p $\sqrt{0.36}$ q $\sqrt{0.5}$ r $\sqrt{0.65}$

Challenge Yourself

1 Which of these are surds? Answer 'Yes' or 'No'.

a $\sqrt[3]{\frac{1}{8}}$ b $\sqrt[4]{\frac{1}{32}}$ c $\sqrt[3]{\frac{64}{125}}$

d $\sqrt[3]{0.125}$ e $\sqrt[3]{0.1}$ f $\sqrt[3]{0.008}$

2 Identify the numbers from the set $\left\{\sqrt{\frac{1}{4}}, \sqrt[3]{\frac{1}{27}}, \sqrt{\frac{1}{8}}, \sqrt{21}, \sqrt[3]{8}\right\}$ that are surds.

Surds: When Close Enough Is Just Not Good Enough

You know what surds are now.

But why use them?

Well, if the root isn't a whole number or a fraction, it'll be a decimal.

And not just any decimal.

It happens to be a never-ending, non-repeating one.

So instead of writing the root as some messy, endless decimal.

We just leave it in surd form.

It's neat.

And easier to work with.

The other reason is accuracy.

Writing out a never-ending decimal?

Not practical.

Round it off, and you're estimating.

Sometimes over. Sometimes under.

But with surds you're using exact values.

You don't lose that accuracy.

This is crucial when you get into higher maths.

Fields like engineering.

Think about designing a bridge.

Or flying a plane.

A tiny bit of rounding can throw everything off.

That's why we use surds.

They're accurate.

Simplifying Surds: No Perfect Squares Allowed

With fractions, you always simplify.

Same goes for surds.

You always simplify them too.

In maths, we try to work with smaller numbers where we can.

Makes calculations shorter and easier.

Before we get into how to simplify surds.

From here on, the surds we'll work with will *only* be square roots.

Because they're the most common ones you'll see.

Also, you'll hear the word "factor" a lot.

A factor is just a number that divides evenly into another number.

To simplify a surd, what's under the square root sign can't have any perfect square factors.

It can have other factors.

Just not perfect square ones.

Remember what a perfect square is?

It's a number you get

when you multiply a number by itself.

In case you forgot, here's a list.

All the squares from 1 to 20.

Keep these in mind.

Now, let's look at how to simplify a surd.

Squared Numbers

$1^2 = 1$	$11^2 = 121$
$2^2 = 4$	$12^2 = 144$
$3^2 = 9$	$13^2 = 169$
$4^2 = 16$	$14^2 = 196$
$5^2 = 25$	$15^2 = 225$
$6^2 = 36$	$16^2 = 256$
$7^2 = 49$	$17^2 = 289$
$8^2 = 64$	$18^2 = 324$
$9^2 = 81$	$19^2 = 361$
$10^2 = 100$	$20^2 = 400$

Here are the rules.

$$\sqrt{n} = \sqrt{a^2 \times b}$$
$$= \sqrt{a^2} \times \sqrt{b}$$
$$= a \times \sqrt{b}$$
$$= a\sqrt{b}$$

$$m\sqrt{n} = m\sqrt{a^2 \times b}$$
$$= m\sqrt{a^2} \times \sqrt{b}$$
$$= ma \times \sqrt{b}$$
$$= ma\sqrt{b}$$

The first rule's saying this.

The number under the square root can be split into two factors.

One of those factors has to be a perfect square.

Now, here's the important part.

That perfect square factor needs to be the **biggest** one.

If it's not, and you go ahead and simplify it.

The surd won't be fully simplified.

Once you've got the two factors.

Split them into separate square roots.

Simplify the perfect square.

Leave the other root as it is.

Then drop the multiplication sign for a cleaner look.

The second rule's pretty much the same thing.

Just with a number outside the root.

Leave it there.

Simplify the square root part.

Then multiply the outside number with the part you simplified.

Here’s another thing.

The number under the square root can have other factors.

Just not perfect square ones.

If it doesn't have any perfect square factors.

Leave it as it is.

It can't be simplified.

Also, it's ok if you don't know what the biggest square factor is right away.

Simplify anyway.

But, double check the final surd.

Make sure it can't be simplified any further.

If it can't.

Sweet.

If it can.

Just repeat the steps.

Eventually, you’ll start noticing the same numbers popping up.

And you’ll know straight away what the biggest square factor is.

Alright, to simplify a surd.

1) Rewrite it as two factors — with one being the biggest perfect square factor
2) Split it into two separate roots
3) Simplify the perfect square, leave the other as is

Worked Example 1

Simplify the following surds.

a $\sqrt{32}$ b $\sqrt{45}$ c $\sqrt{84}$

	Explanation	**Solution**
a	Write the expression.	$\sqrt{32}$
	Find biggest perfect square factor.	$=\sqrt{16\times 2}$
	Split into two square roots.	$=\sqrt{16}\times\sqrt{2}$
	Simplify the perfect square and multiply.	$=4\sqrt{2}$
b	Write the expression.	$\sqrt{45}$
	Find biggest perfect square factor.	$=\sqrt{9\times 5}$
	Split into two square roots.	$=\sqrt{9}\times\sqrt{5}$
	Simplify the perfect square and multiply.	$=3\sqrt{5}$
c	Write the expression.	$\sqrt{84}$
	Find biggest perfect square factor.	$=\sqrt{4\times 21}$
	Split into two square roots.	$=\sqrt{4}\times\sqrt{21}$
	Simplify the perfect square and multiply.	$=2\sqrt{21}$

Worked Example 2

Simplify the following surds.

a $\sqrt{128}$ b $\sqrt{300}$ c $\sqrt{405}$

Explanation		**Solution**
a	Write the expression.	$\sqrt{128}$
	Find biggest perfect square factor.	$= \sqrt{64 \times 2}$
	Split into two square roots.	$= \sqrt{64} \times \sqrt{2}$
	Simplify the perfect square and multiply.	$= 8\sqrt{2}$
b	Write the expression.	$\sqrt{300}$
	Find biggest perfect square factor.	$= \sqrt{100 \times 3}$
	Split into two square roots.	$= \sqrt{100} \times \sqrt{3}$
	Simplify the perfect square and multiply.	$= 10\sqrt{3}$
c	Write the expression.	$\sqrt{405}$
	Find biggest perfect square factor.	$= \sqrt{81 \times 5}$
	Split into two square roots.	$= \sqrt{81} \times \sqrt{5}$
	Simplify the perfect square and multiply.	$= 9\sqrt{5}$

Worked Example 3

Simplify the following.

a $2\sqrt{75}$ b $4\sqrt{48}$ c $3\sqrt{63}$

Explanation	Solution
a Write the expression.	$2\sqrt{75}$
Find biggest perfect square factor.	$= 2\sqrt{25 \times 3}$
Split into two square roots.	$= 2\sqrt{25} \times \sqrt{3}$
Simplify perfect square part and multiply numbers.	$= 2 \times 5 \times \sqrt{3}$
	$= 10\sqrt{3}$
b Write the expression.	$4\sqrt{48}$
Find biggest perfect square factor.	$= 4\sqrt{16 \times 3}$
Split into two square roots.	$= 4\sqrt{16} \times \sqrt{3}$
Simplify perfect square part and multiply numbers.	$= 4 \times 4 \times \sqrt{3}$
	$= 16\sqrt{3}$
c Write the expression.	$3\sqrt{63}$
Find biggest perfect square factor.	$= 3\sqrt{9 \times 7}$
Split into two square roots.	$= 3\sqrt{9} \times \sqrt{7}$
Simplify perfect square part and multiply numbers.	$= 3 \times 3 \times \sqrt{7}$
	$= 9\sqrt{7}$

Worked Example 4

Simplify the following.

a $-7\sqrt{80}$

b $\frac{1}{9}\sqrt{162}$

Explanation	Solution
a Write the expression.	$-7\sqrt{80}$
Find biggest perfect square factor.	$= -7\sqrt{16 \times 5}$
Split into two square roots.	$= -7\sqrt{16} \times \sqrt{5}$
Simplify perfect square part and multiply numbers.	$= -7 \times 4 \times \sqrt{5}$
	$= -28\sqrt{5}$
b Write the expression.	$\frac{1}{9}\sqrt{162}$
Find biggest perfect square factor.	$= \frac{1}{9}\sqrt{81 \times 2}$
Split into two square roots.	$= \frac{1}{9}\sqrt{81} \times \sqrt{2}$
Simplify perfect square part and multiply numbers.	$= \frac{1}{9} \times 9 \times \sqrt{2}$
	$= \sqrt{2}$

Worked Example 5

Express these surds as a square root of a positive integer.

a $3\sqrt{5}$ b $2\sqrt{10}$ c $4\sqrt{6}$

Explanation	Solution
a Write the expression.	$3\sqrt{5}$
Square the outside number and move it inside the square root.	$= \sqrt{3^2 \times 5}$
Multiply numbers inside the square root.	$= \sqrt{9 \times 5}$
	$= \sqrt{45}$
b Write the expression.	$2\sqrt{10}$
Square the outside number and move it inside the square root.	$= \sqrt{2^2 \times 10}$
Multiply numbers inside the square root.	$= \sqrt{4 \times 10}$
	$= \sqrt{40}$
c Write the expression.	$4\sqrt{6}$
Square the outside number and move it inside the square root.	$= \sqrt{4^2 \times 6}$
Multiply numbers inside the square root.	$= \sqrt{16 \times 6}$
	$= \sqrt{96}$

Your Turn To Try

1 **Which of the following is equal to $\sqrt{50}$?**

A $5\sqrt{2}$ B $2\sqrt{5}$

C $2\sqrt{25}$ D $5\sqrt{10}$

2 **What is $\sqrt{200}$ in its simplest form?**

A $100\sqrt{2}$ B $2\sqrt{10}$

C $10\sqrt{2}$ D $4\sqrt{50}$

3 **When simplified $3\sqrt{27}$ is?**

A $3\sqrt{3}$ B $6\sqrt{3}$

C $6\sqrt{3}$ D $9\sqrt{3}$

4 **$\frac{3}{4}\sqrt{80}$ in its simplest for is?**

A $2\sqrt{20}$ B $3\sqrt{5}$

C $4\sqrt{5}$ D $5\sqrt{3}$

5 **Square root of which number is equivalent to $6\sqrt{3}$?**

A 108 B 18

C 54 D 36

Question	1	2	3	4	5
Answer	A	C	D	B	A

Time To Practice

1 Simplify the following surds.

a $\sqrt{12}$ b $\sqrt{18}$ c $\sqrt{20}$

d $\sqrt{24}$ e $\sqrt{27}$ f $\sqrt{72}$

g $\sqrt{98}$ h $\sqrt{125}$ i $\sqrt{360}$

2 Simplify the following surds.

a $2\sqrt{18}$ b $3\sqrt{20}$ c $2\sqrt{63}$

d $4\sqrt{48}$ e $5\sqrt{98}$ f $2\sqrt{800}$

g $4\sqrt{150}$ h $3\sqrt{147}$ i $6\sqrt{75}$

3 Simplify the following surds.

a $\frac{1}{2}\sqrt{20}$ b $\frac{1}{3}\sqrt{28}$ c $\frac{1}{4}\sqrt{54}$

d $\frac{1}{6}\sqrt{24}$ e $\frac{1}{5}\sqrt{80}$ f $\frac{1}{9}\sqrt{99}$

g $\frac{1}{8}\sqrt{84}$ h $\frac{1}{7}\sqrt{56}$ i $\frac{1}{3}\sqrt{68}$

4 Express these surds as a square root of a number.

a $5\sqrt{2}$ b $3\sqrt{3}$ c $8\sqrt{2}$

d $10\sqrt{7}$ e $9\sqrt{10}$ f $3\sqrt{11}$

Challenge Yourself

1 $\sqrt{432}$ in its simplest form is $12\sqrt{k}$. What is the value of k?

2 $\sqrt{450}$ in its simplest form is $c\sqrt{3}$. What is the value of c?

3 $\sqrt{1200}$ can be written in the form $a\sqrt{b}$. What are the value of a and b?

4 Simplify the following surds.

a $\frac{2}{7}\sqrt{98}$

b $\frac{1}{9}\sqrt{162}$

c $\frac{3}{10}\sqrt{175}$

d $\frac{3}{14}\sqrt{147}$

Square and Square Roots: And, We're Back Where We Started

Simplifying surds - done.

Next.

Squaring a square root.

And taking the square root of a perfect square.

These are opposite operations.

Squaring means multiplying a number by itself.

Taking the square root means what number times itself equals what's under the root.

When used together, squaring and taking the square root cancel each other out.

It's like adding, then subtracting the same number.

Or multiplying, then dividing.

They cancel each other out.

Same thing if you reverse it.

Subtracting then adding, or dividing then multiplying.

You end up back where you started.

If you square a number, then take the square root.

Or if you take the square root then square it.

They cancel each other out.

You end up back where you started.

Here's the general rule.

Given $a \geq 0$,

$$\left(\sqrt{a^2}\right) = a$$

$$\left(\sqrt{a}\right)^2 = a$$

Easiest way to remember this.

Just cross out the root sign and the power of 2.

What's left is the answer.

That's the basics.

But questions aren't ever this simple.

So here are three more rules you need to know.

Ones you'll actually use.

If there's something in front of the root — a number or a letter.

And you're squaring the whole thing.

You can't just cross out the root sign and the 2.

Because now, it's not just the square root being squared.

Whatever's in front of the root also gets squared.

For these, use Exponent Law 5.

Square each part separately.

Whatever's in front of the root sign.

And the square root itself.

Given $a \geq 0$,

$$\left(m\sqrt{a}\right)^2 = m^2 \times \left(\sqrt{a}\right)^2$$
$$= m^2 a$$

You can only cross out the root sign and the 2

if the square root is the *only* thing being squared.

The other rule is when you have the square root of two or more perfect square terms.

These can be simplified in two ways.

The first way.

Given $a \geq 0$ and $b \geq 0$,

$$\sqrt{a^2 b^2} = \sqrt{a^2} \times \sqrt{b^2}$$
$$= ab$$

Split each squared term into its own square root.

All being multiplied together.

Cross out the root sign and the 2 for each.

Then multiply what's left to simplify.

The second way.

Use Exponent Law 5 in reverse.

Group the bases of the squared terms inside one bracket.

Then square the whole bracket.

Now you've got the square root of something squared.

Cross out the root sign and the 2.

What's inside the bracket is your answer.

Given $a \geq 0$ and $b \geq 0$,

$$\sqrt{a^2b^2} = \sqrt{(ab)^2}$$
$$= ab$$

Oh, one more thing.

This crossing out of the root sign and the 2.

It only works when it's a square root and a power of 2.

You can't do this with any other powers.

What happens if the power isn't 2?

We'll get to that soon.

For now, just focus on understanding what's happening here.

Because we'll build on this when working with other powers later.

Alright, now let's see what to do when there's a number under the root sign as well.

If the number is a perfect square, follow the same rules.

Using the first rule.

Split the number and each squared term into their own square root.

All being multiplied.

Simplify the number and each square root separately.

Then multiply what's left and combine everything together.

Using the second rule.

Rewrite the number as something squared.

Then group the bases of the squared terms inside one bracket.

Square the whole bracket.

Now you've got the square root of something squared.

Cross out the root sign and the 2.

What's inside the bracket is your answer.

What if the number inside the root is not a perfect square?

Go back to how you simplified surds earlier.

If you take the square root of the number, can it be simplified?

If not, leave it under the root sign.

Simplify anything else.

If it can be simplified.

You can approach it in two ways.

Either.

Split the square root into smaller square roots.

One for the number and one for each term.

Simplify the number and each square root separately.

Then combine everything together.

The final answer will include a surd.

Or.

Keep the number inside the root.

Rewrite it as two numbers multiplied.

Where one number is the biggest perfect square factor.

Write that square factor as something squared.

Then group the bases of the squared terms inside one bracket.

Split the number and the squared term into their own square root.

Simplify each and combine.

The final answer will include a surd.

All this is basically just applying the same two rules.

We're just including numbers now.

By the way.

When we go over the rules where the power isn't 2.

You deal with the numbers exactly how it's explained here.

Cool.

Now, let's look at some examples to make sure you get it.

Worked Example 1

Simplify each of the following:

a $\left(\sqrt{2}\right)^2$ b $\left(\sqrt{12}\right)^2$ c $\left(4\sqrt{5}\right)^2$ d $\left(2\sqrt{7}\right)^2$

Explanation	Solution
a Write the expression.	$\left(\sqrt{2}\right)^2$
Cross out the root sign and 2.	$= 2$
b Write the expression.	$\left(\sqrt{12}\right)^2$
Cross out the root sign and 2.	$= 12$
c Write the expression.	$\left(4\sqrt{5}\right)^2$
Square both number and square root separately.	$= 4^2 \times \left(\sqrt{5}\right)^2$
Simplify.	$= 16 \times 5$
	$= 80$
d Write the expression.	$\left(2\sqrt{7}\right)^2$
Square both number and square root separately.	$= (2)^2 \times \left(\sqrt{7}\right)^2$
Simplify.	$= 4 \times 7$
	$= 28$

Worked Example 2

Simplify the following.

a $\left(\sqrt{5x}\right)^2$ b $\left(\sqrt{7x^2}\right)^2$ c $\left(-2\sqrt{3x}\right)^2$

Explanation **Solution**

a Write the expression. $\left(\sqrt{5x}\right)^2$

Cross out the root sign and 2. $= 5x$

b Write the expression. $\left(\sqrt{7x^2}\right)^2$

Cross out the root sign and 2. $= 7x^2$

c Write the expression. $\left(-2\sqrt{3x}\right)^2$

Square both number and square root separately. $= (-2)^2 \times \left(\sqrt{3x}\right)^2$

Simplify. $= 4 \times 3x$

$= 12x$

When squaring a square root, put the whole root in brackets. That way, it's clear you're squaring the entire root, not just what's inside.

Worked Example 3

Simplify each of the following, where $a \geq 0$ and $b \geq 0$:

a $\sqrt{3^2}$ b $\sqrt{16a^2}$ c $3\sqrt{49a^2b^2}$

Explanation	Solution
a Write the expression.	$\sqrt{3^2}$
Cross out the root sign and 2.	$= 3$
b Write the expression.	$\sqrt{16a^2}$
Split into separate square roots.	$= \sqrt{16} \times \sqrt{a^2}$
Simplify square roots and combine.	$= 4a$
	or
Rewrite as a squared term.	$= \sqrt{(4a)^2}$
Cross out the root sign and 2.	$= 4a$
c Write the expression.	$3\sqrt{49a^2b^2}$
Split into separate square roots.	$= 3\sqrt{49} \times \sqrt{a^2} \times \sqrt{b^2}$
Simplify square roots.	$= 3 \times 7 \times a \times b$
Combine.	$= 21ab$
	or
Rewrite as a squared term.	$= 3\sqrt{(7ab)^2}$
Cross out the root sign and 2.	$= 3 \times 7ab$
	$= 21ab$

Worked Example 4

Simplify each of the following, where $x \geq 0$ *and* $y \geq 0$:

a $\sqrt{6x^2}$ b $\sqrt{27y^2}$

Explanation	Solution
a Write the expression.	$\sqrt{6x^2}$
Cross out the root sign and 2.	$= \sqrt{6} \times \sqrt{x^2}$
	$= x\sqrt{6}$
b Write the expression.	$\sqrt{27y^2}$
Split number and each squared term.	$= \sqrt{27} \times \sqrt{y^2}$
Simplify the number part and square root.	$= \sqrt{9 \times 3} \times y$
	$= 3\sqrt{3} \times y$
Combine.	$= 3y\sqrt{3}$
	or
Write the expression.	$\sqrt{27y^2}$
Rewrite number as two numbers multiplied.	$= \sqrt{9 \times 3 \times y^2}$
Write square factor as something squared.	$= \sqrt{3^2 \times 3 \times y^2}$
Group bases of squared terms inside one bracket.	$= \sqrt{(3y)^2 \times 3}$
Split into two square roots.	$= \sqrt{(3y)^2} \times \sqrt{3}$
Simplify each and combine.	$= 3y\sqrt{3}$

Worked Example 5

Simplify in surd form, where a and b are positive numbers. $\frac{1}{2}\sqrt{128a^2b^2}$

Explanation	Solution
Write the expression.	$\frac{1}{2}\sqrt{128a^2b^2}$
Split number part and each squared term.	$= \frac{1}{2}\sqrt{128} \times \sqrt{a^2} \times \sqrt{b^2}$
Simplify the number and square roots.	$= \frac{1}{2}\sqrt{64 \times 2} \times a \times b$
Combine.	$= \frac{1}{2} \times 8\sqrt{2} \times ab$
	$= 4ab\sqrt{2}$
	or
Write the expression.	$\frac{1}{2}\sqrt{128a^2b^2}$
Rewrite number part as two numbers multiplied.	$= \frac{1}{2}\sqrt{64 \times 2 \times a^2b^2}$
Write square factor as something squared.	$= \frac{1}{2}\sqrt{8^2 \times 2 \times a^2b^2}$
Group bases of squared terms inside one bracket.	$= \frac{1}{2}\sqrt{(8ab)^2 \times 2}$
Split into two square roots.	$= \frac{1}{2}\sqrt{(8ab)^2} \times \sqrt{2}$
Simplify and combine.	$= \frac{1}{2} \times 8ab\sqrt{2} = 4ab\sqrt{2}$

When Powers Aren't 2: For Even Powers, Just Go Halfsies

That was a couple of examples on squares and square roots.

Everything had a power of 2.

Not too bad.

But what if the power isn't 2?

What then?

Well, it depends if the power is even or odd.

Even powers are easier.

So we'll start with those.

Once you know what to do with them, odd powers will make sense too.

Before we get into it.

Think back to exponent Laws 4 and 5.

The ones where there's a power outside the brackets.

We're gonna use them, but in reverse.

Here's the general rule.

Given m is an even number.

$$\sqrt{a^m} = \sqrt{\left(a^{\frac{m}{2}}\right)^2}$$

$$= a^{\frac{m}{2}}$$

For even powers.

Halve the power.

Write the base with that halved power inside brackets.

Then square the whole bracket.

Now you have a square root of something squared.

Cross out the root sign and the 2.

What's left is the answer.

If there's more than one letter raised to an even power.

Halve each power.

Write each letter with its halved power inside a bracket.

Square the brackets.

Then group the bases of the squared terms inside one bracket.

Now you've got the square root of something squared.

Cross out the root sign and the 2.

What's inside the bracket is your answer.

Given m and n are even numbers.

$$\sqrt{a^m b^n} = \sqrt{\left(a^{\frac{m}{2}}\right)^2 \left(b^{\frac{n}{2}}\right)^2}$$

$$= \sqrt{\left(a^{\frac{m}{2}} b^{\frac{n}{2}}\right)^2}$$

$$= a^{\frac{m}{2}} b^{\frac{n}{2}}$$

This is the faster way.

But if you feel more comfortable

splitting each term into separate roots.

You can do that too.

In the examples, you'll see both ways.

Just so you get a good feel for each.

If there are numbers inside the root.

Follow the same steps explained before.

Worked Example 6

Simplify each of the following, assume a and b are positive numbers.

a $\sqrt{a^6}$ b $\sqrt{b^8}$ c $\sqrt{4a^4}$

Explanation		**Solution**
a	Write the expression.	$\sqrt{a^6}$
	Halve the power and rewrite as a squared term.	$=\sqrt{(a^3)^2}$
	Simplify.	$=a^3$
b	Write the expression.	$\sqrt{b^8}$
	Halve the power and rewrite as a squared term.	$=\sqrt{(b^4)^2}$
	Simplify.	$=b^4$
c	Write the expression.	$\sqrt{4a^4}$
	Split into two square roots.	$=\sqrt{4}\times\sqrt{a^4}$
	Rewrite second root as a squared term.	$=2\times\sqrt{(a^2)^2}$
	Simplify	$=2a^2$
		or
	Write the expression.	$\sqrt{4a^4}$
	Rewrite number and term as something squared. Group bases inside one bracket.	$=\sqrt{2^2(a^2)^2}=\sqrt{(2a^2)^2}$
	Simplify.	$=2a^2$

Worked Example 7

Simplify each of the following, assume x and y are positive numbers.

a $\sqrt{25x^4y^2}$ b $\sqrt{64x^8y^6}$ c $4\sqrt{36x^{14}y^{10}}$

Explanation	Solution
a Write the expression.	$\sqrt{25x^4y^2}$
Split into three square roots.	$= \sqrt{25} \times \sqrt{x^4} \times \sqrt{y^2}$
Rewrite middle root as a squared term.	$= 5 \times \sqrt{(x^2)^2} \times y$
Simplify and combine the results.	$= 5 \times x^2 \times y$
	$= 5x^2y$
b Write the expression.	$\sqrt{64x^8y^6}$
Split into three square roots.	$= \sqrt{64} \times \sqrt{x^8} \times \sqrt{y^6}$
Rewrite two roots as a squared term.	$= 8 \times \sqrt{(x^4)^2} \times \sqrt{(y^3)^2}$
Simplify and combine the results.	$= 8 \times x^4 \times y^3$
	$= 8x^4y^3$
c Write the expression.	$4\sqrt{36x^{14}y^{10}}$
Split into three square roots.	$= 4\sqrt{36} \times \sqrt{x^{14}} \times \sqrt{y^{10}}$
Rewrite two roots as a squared term.	$= 4 \times 6 \times \sqrt{(x^7)^2} \times \sqrt{(y^5)^2}$
Simplify and combine the results.	$= 24 \times x^7 \times y^5$
	$= 24x^7y^5$

Worked Example 8

Simplify each of the following, assume x and y are positive numbers.

a $\sqrt{25x^4y^2}$ b $\sqrt{64x^8y^6}$ c $4\sqrt{36x^{14}y^{10}}$

Explanation	Solution
a Write the expression.	$\sqrt{25x^4y^2}$
Rewrite the number as something squared.	$= \sqrt{5^2x^4y^2}$
Write each term with its halved power inside a bracket.	$= \sqrt{(5x^2y)^2}$
Simplify.	$= 5x^2y$
b Write the expression.	$\sqrt{64x^8y^6}$
Rewrite the number as something squared.	$= \sqrt{8^2x^8y^6}$
Write each term with its halved power inside a bracket.	$= \sqrt{(8x^4y^3)^2}$
Simplify.	$= 8x^4y^3$
c Write the expression.	$4\sqrt{36x^{14}y^{10}}$
Rewrite the number as something squared.	$= 4\sqrt{6^2x^{14}y^{10}}$
Write each term with its halved power inside a bracket.	$= 4\sqrt{(6x^7y^5)^2}$
Simplify and multiply by outside number.	$= 4 \times 6x^7y^5 = 24x^7y^5$

Worked Example 9

Express in simplest form where a and b are positive integers.

a $\sqrt{18a^4}$

b $5\sqrt{20a^8b^6}$

Explanation	Solution
a Write the expression.	$\sqrt{18a^4}$
Rewrite number as two numbers multiplied.	$= \sqrt{9a^4 \times 2}$
Write square factor and even powered term as something squared.	$= \sqrt{(3a^2)^2 \times 2}$
Split into separate square roots.	$= \sqrt{(3a^2)^2} \times \sqrt{2}$
Simplify.	$= 3a^2\sqrt{2}$
b Write the expression.	$5\sqrt{20a^8b^6}$
Rewrite number inside root as two numbers multiplied.	$= 5\sqrt{4a^8b^6 \times 5}$
Write square factor and even powered term as something squared.	$= 5\sqrt{(2a^4b^3)^2 \times 5}$
Split into separate square roots.	$= 5 \times \sqrt{(2a^4b^3)^2} \times \sqrt{5}$
Simplify.	$= 5 \times 2a^4b^3 \times \sqrt{5}$
	$= 10a^4b^3\sqrt{5}$

When Powers Aren't 2: For Odd Powers, Take One Then Go Halfsies

Now we'll look at what to do with odd powers.

For this, think back to Exponent Law 2.

The one where you add the powers.

Because we're going to use that idea here.

But in reverse.

We're going to rewrite one term as two terms.

Here's the general rule.

Given $a \geq 0$

and $'m'$ is an odd number

$$\sqrt{a^m} = \sqrt{a^{m-1} \times a}$$

$$= \sqrt{\left(a^{\frac{m-1}{2}}\right)^2 \times a}$$

$$= \sqrt{\left(a^{\frac{m-1}{2}}\right)^2} \times \sqrt{a}$$

$$= a^{\frac{m-1}{2}} \sqrt{a}$$

When you subtract 1 from an odd number, you get an even number.

So, with an odd powered term.

You split it into two parts.

One with an even power.

The original odd power minus 1.

The other just the letter by itself.

Which really has a power of 1.

For the even powered term.

Halve the power, put it in brackets, and square the whole bracket.

From there, split the square root into two separate roots.

Simplify the one with the even power.

Just like before.

Keep the other root as is.

And simplify.

If there's more than one term with odd powers.

Same thing.

Split each into two terms.

One having an even power.

The original odd power minus 1.

The other is just the letter by itself.

Given $a \geq 0$,

'm' and 'n' are odd

$$\sqrt{a^m b^n} = \sqrt{a^{m-1} b^{n-1} \times ab}$$

$$= \sqrt{\left(a^{\frac{m-1}{2}} b^{\frac{n-1}{2}}\right)^2 \times ab}$$

$$= \sqrt{\left(a^{\frac{m-1}{2}} b^{\frac{n-1}{2}}\right)^2} \times \sqrt{ab}$$

$$= a^{\frac{m-1}{2}} b^{\frac{n-1}{2}} \sqrt{ab}$$

From here, you'll have even powered terms

and single letters.

Simplify the even-powered ones as before.

The single letters will stay inside the root.

If there's a mix of even and odd powers.

Split the odd powered terms into two.

Combine all the even-powered terms into one squared term multiplied by what's left.

Simplify the same way as before.

If there are numbers as well, simplify them the same way you've already learned.

Alright, let's go through a few examples to see how this works

Worked Example 10

Simplify each of the following in surd form, assume x and y are positive numbers.

a $\sqrt{x^5}$ b $\sqrt{y^7}$

Explanation	**Solution**
a Write the expression.	$\sqrt{x^5}$
Subtract 1 from power and split into two parts.	$= \sqrt{x^4 \times x}$
Rewrite as a squared term multiplied by a single term.	$= \sqrt{(x^2)^2 \times x}$
Split into separate square roots.	$= \sqrt{(x^2)^2} \times \sqrt{x}$
Simplify.	$= x^2\sqrt{x}$
b Write the expression.	$\sqrt{y^7}$
Subtract 1 from power and split into two parts.	$= \sqrt{y^6 \times y}$
Rewrite as a squared term multiplied by a single term.	$= \sqrt{(y^3)^2 \times y}$
Split into separate square roots.	$= \sqrt{(y^3)^2} \times \sqrt{y}$
Simplify.	$= y^3\sqrt{y}$

Worked Example 11

Simplify each of the following in surd form, assume x and y are positive numbers.

a $2\sqrt{9x^3}$ b $4\sqrt{50y^7}$

Explanation	Solution
a Write the expression.	$2\sqrt{9x^3}$
Subtract 1 from power and split letter into two parts.	$= 2\sqrt{9x^2 \times x}$
Rewrite as a squared term multiplied by what's left.	$= 2\sqrt{(3x)^2 \times x}$
Split into separate square roots.	$= 2\sqrt{(3x)^2} \times \sqrt{x}$
Simplify.	$= 2 \times 3x \times \sqrt{x}$
	$= 6x\sqrt{x}$
b Write the expression.	$4\sqrt{50y^7}$
Subtract 1 from power and split letter into two parts. Rewrite number as two factors.	$= 4\sqrt{25 \times 2 \times y^6 \times y}$
Group perfect square factor and even powers together.	$= 4\sqrt{25y^6 \times 2y}$
Rewrite as a squared term multiplied by what's left.	$= 4\sqrt{(5y^3)^2 \times 2y}$
Split into separate square roots.	$= 4\sqrt{(5y^3)^2} \times \sqrt{2y}$
Multiply by number outside and simplify.	$= 4 \times 5y^3 \times \sqrt{2y}$
	$= 20y^3\sqrt{2y}$

Worked Example 12

Simplify each of the following in surd form, assume a and b are positive numbers.

a $\sqrt{a^5b^3}$

b $\sqrt{15a^9b^7}$

Explanation	Solution
a Write the expression.	$\sqrt{a^5b^3}$
Subtract 1 from each power. Split into two parts.	$= \sqrt{a^4b^2 \times ab}$
Rewrite as a squared term multiplied by what's left.	$= \sqrt{(a^2b)^2 \times ab}$
Split into separate square roots.	$= \sqrt{(a^2b)^2} \times \sqrt{ab}$
Simplify.	$= a^2b\sqrt{ab}$
b Write the expression.	$\sqrt{15a^9b^7}$
Subtract 1 from each power. Split into two parts.	$= \sqrt{a^8b^6 \times 15ab}$
Rewrite as a squared term multiplied by what's left.	$= \sqrt{(a^4b^3)^2 \times 15ab}$
Split into separate square roots.	$= \sqrt{(a^4b^3)^2} \times \sqrt{15ab}$
Simplify.	$= a^4b^3\sqrt{15ab}$

Simplify all numbers and letters. Make sure there are no perfect square factors left inside the root sign

Worked Example 13

Simplify in surd form, where a and b are positive numbers. $\frac{1}{2}\sqrt{27a^3b^{11}}$

Explanation	**Solution**
Write the expression.	$\frac{1}{2}\sqrt{27a^3b^{11}}$
Subtract 1 from each power and split the letters. Rewrite the number as two factors.	$=\frac{1}{2}\sqrt{9\times 3\times a^2\,b^{10}\times ab}$
Group perfect square factor and even powers together.	$=\frac{1}{2}\sqrt{9a^2b^{10}\times 3ab}$
Rewrite as a squared part multiplied by what's left.	$=\frac{1}{2}\sqrt{(3ab^5)^2\times 3ab}$
Split into separate square roots.	$=\frac{1}{2}\sqrt{(3ab^5)^2}\times\sqrt{3ab}$
Multiply by number outside and simplify.	$=\frac{1}{2}\times 3ab^5\times\sqrt{3ab}$
	$=\frac{3}{2}ab^5\sqrt{3ab}$

Not grouping terms with different powers correctly,
they should be written as one squared part multiplied by what's left.

Worked Example 14

Simplify in surd form, where r and s are positive numbers. $\frac{3}{4}\sqrt{80r^{10}s^3}$

Explanation	**Solution**
Write the expression.	$\frac{3}{4}\sqrt{80r^{10}s^3}$
Subtract 1 from odd power and split the letter. Rewrite the number as two factors.	$=\frac{3}{4}\sqrt{16\times 5\times r^{10}\times s^2\times s}$
Group perfect square factor and even powers together.	$=\frac{3}{4}\sqrt{16r^{10}s^2\times 5s}$
Rewrite as a squared part multiplied by what's left.	$=\frac{3}{4}\sqrt{(4r^5s)^2\times 5s}$
Split into separate square roots.	$=\frac{3}{4}\sqrt{(4r^5s)^2}\times\sqrt{5s}$
Multiply by number outside and simplify.	$=\frac{3}{4}\times 4r^5s\times\sqrt{5s}$
	$=3r^5s\sqrt{5s}$

Your Turn To Try

1 **What is $(\sqrt{5})^2$?**

A 25 B 5

C $2\sqrt{5}$ D $\sqrt{10}$

2 **What is $(2\sqrt{4x})^2$ in its simplest form?**

A $8x$ B $8x^2$

C $16x^2$ D $16x$

3 **When simplified $\sqrt{81a^4}$ is?**

A $9a^4$ B $9a^3$

C $9a^2$ D $9a$

4 **$\sqrt{100x^6y^4}$ in its simplest for is?**

A $10x^6y^4$ B $10x^4y^6$

C $10x^3y^2$ D $10x^2y^3$

5 **Which surd is equivalent to $\sqrt{200x^5}$?**

A $10x^2\sqrt{2x}$ B $10x^2\sqrt{5}$

C $20x^3\sqrt{2}$ D $10x^2\sqrt{2}$

Question	1	2	3	4	5
Answer	B	D	C	C	A

Time To Practice

1 Simplify the following surds.

a $(\sqrt{7})^2$ b $(\sqrt{11})^2$ c $(\sqrt{13})^2$

d $(\sqrt{5x})^2$ e $(\sqrt{9x})^2$ f $(\sqrt{3x^2})^2$

g $(2\sqrt{2x})^2$ h $(-4\sqrt{3x})^2$ i $(-5\sqrt{2x^2})^2$

2 Simplify the following surds.

a $\sqrt{z^2}$ b $\sqrt{25z^2}$ c $\sqrt{121z^2}$

d $\sqrt{36b^2}$ e $\sqrt{100b^2}$ f $\sqrt{49b^2}$

g $\sqrt{144p^2q^2}$ h $\sqrt{225p^6q^4}$ i $\sqrt{289p^2q^8}$

3 Simplify the following surds, where x and y are positive numbers.

a $\sqrt{12x^2}$ b $\sqrt{28x^2}$ c $\sqrt{45y^2}$

d $\sqrt{48x^8}$ e $\sqrt{50x^6}$ f $\sqrt{72y^4}$

g $\sqrt{75x^8y^6}$ h $\sqrt{96x^4y^2}$ i $\sqrt{98x^2y^6}$

4 Simplify the following in surd form, where a and b are positive numbers.

a $\sqrt{a^{13}}$ b $\sqrt{2b^5}$ c $\sqrt{4a^7}$

d $3\sqrt{16a^9}$ e $2\sqrt{4a^3b}$ f $\sqrt{45a^7b^5}$

g $3\sqrt{14ab^5}$ h $4\sqrt{18a^{15}b^3}$ i $5\sqrt{15a^5b^9}$

Challenge Yourself

1 Simplify the following surds, where x and y are positive numbers.

a $3\sqrt{125x^4y^4}$

b $7\sqrt{162x^2y^6}$

c $\frac{1}{2}\sqrt{88x^4y^{12}}$

d $\frac{3}{5}\sqrt{80x^{18}y^{16}}$

e $\frac{1}{15}\sqrt{75x^4y^6z^8}$

f $\frac{2}{3}\sqrt{54x^8y^2z^4}$

g $\frac{3}{2}\sqrt{48x^9y^3z^5}$

h $\frac{5}{2}\sqrt{108x^7yz^5}$

Adding and Subtracting Surds: This is for Look-Alikes Only

We've only looked at simplifying surds so far.

But that's not all you can do with them.

You can also add, subtract, multiply, and divide surds.

We'll go over how each one works.

Starting with adding and subtracting.

Here are the rules.

Surd Rules

Addition	$m\sqrt{a} + n\sqrt{a} = (m + n)\sqrt{a}$
Subtraction	$m\sqrt{a} - n\sqrt{a} = (m - n)\sqrt{a}$

If the surds are the same, meaning what's under the root sign is identical.

Just add or subtract the numbers in front.

The surd part stays the same.

Simple.

If the surds are different.

You can't add or subtract them.

Just leave them.

You can only add or subtract if the surds are exactly the same.

However.

Just because they look different, it doesn’t always mean they are.

It could be that the surds aren't in their simplest form.

So, before you add or subtract.

Double check if the surds are in simplest form.

Sometimes, it's only after simplifying that you notice they're actually the same.

But, if the surds are different and you can't simplify them any further.

Just leave them as they are.

You can’t add or subtract them.

Worked Example 1

Evaluate the following.

a $\sqrt{3} + \sqrt{3}$

b $4\sqrt{2} + 3\sqrt{2}$

c $\sqrt{5} + 2\sqrt{5} + 3\sqrt{5}$

Explanation	Solution
a Write the expression.	$\sqrt{3} + \sqrt{3}$
Combine like surds by adding.	$= (1 + 1)\sqrt{3}$
Simplify.	$= 2\sqrt{3}$
b Write the expression.	$4\sqrt{2} + 3\sqrt{2}$
Combine like surds by adding.	$= (4 + 3)\sqrt{2}$
Simplify.	$= 7\sqrt{2}$
c Write the expression.	$\sqrt{5} + 2\sqrt{5} + 3\sqrt{5}$
Combine like surds by adding.	$= (1 + 2 + 3)\sqrt{5}$
Simplify.	$= 6\sqrt{5}$

Worked Example 2

Evaluate the following.

a $5\sqrt{7} - \sqrt{7}$ b $8\sqrt{10} - 3\sqrt{10}$ c $12\sqrt{11} - 3\sqrt{11} - 4\sqrt{11}$

Explanation		Solution
a	Write the expression.	$5\sqrt{7} - \sqrt{7}$
	Combine like surds by subtracting.	$= (5 - 1)\sqrt{7}$
	Simplify.	$= 4\sqrt{7}$
b	Write the expression.	$8\sqrt{10} - 3\sqrt{10}$
	Combine like surds by subtracting.	$= (8 - 3)\sqrt{10}$
	Simplify.	$= 5\sqrt{10}$
c	Write the expression.	$12\sqrt{11} - 3\sqrt{11} - 4\sqrt{11}$
	Combine like surds by subtracting.	$= (12 - 3 - 4)\sqrt{11}$
	Simplify.	$= 5\sqrt{11}$

When adding or subtracting surds, you don't have to use brackets.
Just add or subtract the numbers in front and leave the surd as is.

Worked Example 3

Evaluate the following.

a $\mathbf{2\sqrt{5} + \sqrt{3} + 4\sqrt{5}}$

b $\mathbf{\sqrt{13} + 7\sqrt{15} - 9\sqrt{15}}$

c $\mathbf{-2\sqrt{7} + \sqrt{3} - 3\sqrt{7} - 10\sqrt{3}}$

d $\mathbf{-4\sqrt{5} - 2\sqrt{15} + 5\sqrt{15} + 2\sqrt{5}}$

Explanation	**Solution**
a Write the expression.	$2\sqrt{5} + \sqrt{3} - \sqrt{5}$
Combine like surds.	$= 6\sqrt{5} + \sqrt{3}$
b Write the expression.	$\sqrt{13} + 7\sqrt{15} - 9\sqrt{15}$
Combine like surds.	$= \sqrt{13} - 2\sqrt{15}$
c Write the expression.	$-2\sqrt{7} + \sqrt{3} - 3\sqrt{7} - 10\sqrt{3}$
Combine like surds.	$= -5\sqrt{7} - 7\sqrt{3}$
d Write the expression.	$-4\sqrt{5} - 2\sqrt{15} + 5\sqrt{15} + 2\sqrt{5}$
Combine like surds.	$= -2\sqrt{5} + 3\sqrt{15}$

Adding or subtracting surds when the numbers or letter under the root signs are different. ⚠

Worked Example 4

Evaluate the following.

a $\sqrt{8}-\sqrt{2}$ b $\sqrt{27}+\sqrt{3}$ c $4\sqrt{18}-5\sqrt{2}$

Explanation	**Solution**
a Write the expression.	$\sqrt{8}-\sqrt{2}$
Find the highest perfect square factor of first surd.	$=\sqrt{4\times 2}-\sqrt{2}$
Split into separate square roots.	$=\sqrt{4}\times\sqrt{2}-\sqrt{2}$
Simplify perfect square.	$=2\sqrt{2}-\sqrt{2}$
Combine like surds.	$=\sqrt{2}$
b Write the expression.	$\sqrt{27}+\sqrt{3}$
Find the highest perfect square factor of first surd.	$=\sqrt{9\times 3}+\sqrt{3}$
Split into separate square roots.	$=\sqrt{9}\times\sqrt{3}+\sqrt{3}$
Simplify perfect square.	$=3\sqrt{3}+\sqrt{3}$
Combine like surds.	$=4\sqrt{3}$
c Write the expression.	$4\sqrt{18}-5\sqrt{2}$
Find the highest perfect square factor of first surd.	$=4\sqrt{9\times 2}-5\sqrt{2}$
Split into separate square roots.	$=4\sqrt{9}\times\sqrt{2}-5\sqrt{2}$
Simplify perfect square.	$=4(3)\times\sqrt{2}-5\sqrt{2}$
Combine like surds.	$=12\sqrt{2}-5\sqrt{2}$
	$=7\sqrt{2}$

Worked Example 5

Simplify the following surds. Assume that x and y are positive real numbers.

a $2\sqrt{x}+5\sqrt{y}+\sqrt{x}+3\sqrt{y}$

b $12\sqrt{x}-8\sqrt{y}-2\sqrt{x}-9\sqrt{y}$

c $\sqrt{3x}+2\sqrt{3x}-\sqrt{6y}$

d $3\sqrt{5x}-4\sqrt{10y}-4\sqrt{5x}-\sqrt{10y}$

Explanation	Solution
a Write the expression.	$2\sqrt{x}+5\sqrt{y}+\sqrt{x}+3\sqrt{y}$
Combine like surds.	$=3\sqrt{x}+8\sqrt{y}$
b Write the expression.	$12\sqrt{x}-8\sqrt{y}-2\sqrt{x}-9\sqrt{y}$
Combine like surds.	$=10\sqrt{x}-17\sqrt{y}$
c Write the expression.	$\sqrt{3x}+2\sqrt{3x}-\sqrt{6y}$
Combine like surds.	$=3\sqrt{3x}-\sqrt{6y}$
d Write the expression.	$3\sqrt{5x}-4\sqrt{10y}-4\sqrt{5x}-\sqrt{10y}$
Combine like surds.	$=-\sqrt{5x}-5\sqrt{10y}$

Not checking if surds can be fully simplified before adding or subtracting. ⚠

Your Turn To Try

1 **When simplified form of $3\sqrt{5} + 2\sqrt{5}$?**

A $5\sqrt{5}$
B $6\sqrt{5}$
C $5\sqrt{10}$
D 25

2 **Find the simplified value of $4\sqrt{7} - \sqrt{7}$.**

A $2\sqrt{7}$
B $5\sqrt{7}$
C $3\sqrt{7}$
D 28

3 **What is the result when you simplify $\sqrt{12} + \sqrt{27}$?**

A $3\sqrt{5}$
B $4\sqrt{3}$
C $4\sqrt{3}$
D $5\sqrt{3}$

4 **What do you get when you simplify $4\sqrt{50} - 5\sqrt{8}$?**

A $10\sqrt{2}$
B $13\sqrt{2}$
C $9\sqrt{2}$
D $7\sqrt{2}$

5 **Which of these is the simplified version of $\sqrt{18x} - 2\sqrt{8x}$**

A $\sqrt{2x}$
B $3\sqrt{2x}$
C $-\sqrt{2x}$
D $-3\sqrt{2}$

Question	1	2	3	4	5
Answer	A	C	D	A	C

Time To Practice

1 Simplify the following.

a $4\sqrt{7} + 3\sqrt{7}$ b $5\sqrt{3} + 8\sqrt{3}$ c $12\sqrt{2} + 3\sqrt{2}$

d $6\sqrt{3} - 2\sqrt{3}$ e $11\sqrt{10} - 6\sqrt{10}$ f $15\sqrt{5} - \sqrt{5}$

g $-2\sqrt{13} - \sqrt{13} + 4\sqrt{13}$ h $-\sqrt{21} + 5\sqrt{21} - 2\sqrt{21}$

2 Simplify the following.

a $8\sqrt{5} + 3\sqrt{3} + 7\sqrt{5} + 2\sqrt{3}$ b $12\sqrt{3} - 8\sqrt{7} + 5\sqrt{3} - 10\sqrt{7}$

c $2\sqrt{2} - 4\sqrt{10} - 5\sqrt{2} + \sqrt{10}$ d $-4\sqrt{11} - 2\sqrt{15} + 7\sqrt{11} + 4\sqrt{15}$

3 Simplify the following.

a $\sqrt{8} + 3\sqrt{2}$ b $\sqrt{20} - \sqrt{5}$ c $2\sqrt{27} + 4\sqrt{3}$

d $2\sqrt{24} + 3\sqrt{6}$ e $3\sqrt{72} + 2\sqrt{98}$ f $2\sqrt{125} - 3\sqrt{45}$

g $2\sqrt{80} - \sqrt{45} + 2\sqrt{63}$ h $\sqrt{36} - \sqrt{108} - 3\sqrt{3}$

4 Simplify the following. Assume that x and y are positive numbers.

a $2\sqrt{x} + 5\sqrt{y} + 6\sqrt{x} - 2\sqrt{y}$ b $7\sqrt{x} - \sqrt{y} + 2\sqrt{9x} - 3\sqrt{y}$

c $10\sqrt{y} - 2\sqrt{4y} + 2\sqrt{x} - \sqrt{x}$ d $6\sqrt{x} - 12\sqrt{9y} - 7\sqrt{x} + 2\sqrt{y}$

Challenge Yourself

1 $4\sqrt{2} + \sqrt{32}$ can be expressed in a single surd in the form $a\sqrt{b}$. What are the values of a and b?

2 Combine $3\sqrt{125} - 2\sqrt{45}$ into the form $a\sqrt{b}$. What are the values of a and b?

3 Simplify $4\sqrt{50x} - 3\sqrt{18x}$ into the form $a\sqrt{bx}$. What are the values of a and b?

4 Simplify the following surds.

a $10\sqrt{3a} - 15\sqrt{27a} + 8\sqrt{12a}$

b $6\sqrt{2b} + \sqrt{128b} - \sqrt{144}$

c $\frac{1}{8}\sqrt{32a} - \frac{7}{6}\sqrt{18a} - \frac{1}{3}\sqrt{72a}$

d $\frac{1}{2}\sqrt{150ab} + \frac{1}{4}\sqrt{96ab} - \frac{1}{6}\sqrt{6ab}$

Multiplying Surds: Look-Alike or Not, Multiply Anyway

Multiplying surds is a bit easier.

You don't have to worry about the surds being the same.

They can be anything - it doesn't matter.

Here's the general rule.

Surd Rules

Multiplication	$\sqrt{a} \times \sqrt{b} = \sqrt{ab}$ $m\sqrt{a} \times n\sqrt{b} = mn\sqrt{ab}$

When multiplying surds, just multiply what's inside the roots.

All under one root.

If there's a number outside the root signs.

Multiply those separately, and keep them outside.

The only thing is.

After multiplying, you might need to simplify the answer.

So, always double check.

And simplify the answer if you can.

If the surds aren't in their simplest form the start.

Simplify them first.

Before you multiply.

That way, you're working with smaller numbers.

After you multiply.

Double check if the answer can be simplified further.

Worked Example 1

Evaluate the following.

a $\sqrt{2} \times \sqrt{5}$ b $\sqrt{6} \times \sqrt{7}$ c $\sqrt{21} \times \sqrt{3}$

Explanation		**Solution**
a	Write the expression.	$\sqrt{2} \times \sqrt{5}$
	Multiply what's inside the roots under one root.	$= \sqrt{2 \times 5}$
	Simplify.	$= \sqrt{10}$
b	Write the expression.	$\sqrt{6} \times \sqrt{7}$
	Multiply what's inside the roots under one root.	$= \sqrt{6 \times 7}$
	Simplify.	$= \sqrt{42}$
c	Write the expression.	$\sqrt{21} \times \sqrt{3}$
	Multiply what's inside the roots under one root.	$= \sqrt{21 \times 3}$
	Simplify.	$= \sqrt{63}$
	Find the largest perfect square factor.	$= \sqrt{9 \times 7}$
	Take out the perfect square and simplify.	$= 3\sqrt{7}$

If you see a perfect square number inside the root, take it out and write its square root in front of the root sign.

Worked Example 2

Evaluate the following.

a $\mathbf{5\sqrt{3} \times 2\sqrt{11}}$ b $\mathbf{10\sqrt{15} \times 6\sqrt{3}}$ c $\mathbf{4\sqrt{10} \times 3\sqrt{5}}$

Explanation	Solution
a Write the expression.	$5\sqrt{3} \times 2\sqrt{11}$
Multiply the numbers outside. Multiply numbers inside under one root.	$= 10\sqrt{3 \times 11}$
Simplify.	$= 10\sqrt{33}$
b Write the expression.	$10\sqrt{15} \times 6\sqrt{3}$
Multiply the numbers outside. Multiply numbers inside under one root.	$= 60\sqrt{45}$
Find the largest perfect square factor.	$= 60\sqrt{9 \times 5}$
Take out the perfect square.	$= 60 \times 3\sqrt{5}$
Simplify.	$= 180\sqrt{5}$
c Write the expression.	$4\sqrt{10} \times 3\sqrt{5}$
Multiply the numbers outside. Multiply numbers inside under one root.	$= 12\sqrt{50}$
Find the largest perfect square factor.	$= 12\sqrt{25 \times 2}$
Take out the perfect square.	$= 12 \times 5\sqrt{2}$
Simplify.	$= 60\sqrt{2}$

Worked Example 3

Evaluate the following.

a $\sqrt{48} \times 2\sqrt{2}$

b $4\sqrt{27} \times 2\sqrt{50}$

Explanation	Solution
a Write the expression.	$\sqrt{48} \times 2\sqrt{2}$
Find the largest perfect square factor of first root.	$= \sqrt{16 \times 3} \times 2\sqrt{2}$
Take out the perfect square.	$= 4\sqrt{3} \times 2\sqrt{2}$
Multiply and simplify.	$= 8\sqrt{6}$
b Write the expression.	$4\sqrt{27} \times 2\sqrt{50}$
Find the largest perfect square factor of each root.	$= 4\sqrt{9 \times 3} \times 2\sqrt{25 \times 2}$
Take out the perfect square from each root.	$= 4 \times 3\sqrt{3} \times 2 \times 5\sqrt{2}$
	$= 12\sqrt{3} \times 10\sqrt{2}$
Multiply the numbers outside. Multiply numbers inside under one root.	$= 120\sqrt{6}$

Multiplying the numbers outside the root sign by the numbers inside it. ⚠

Worked Example 4

Simplify the following surds involving fractions.

a $\frac{\sqrt{3}}{2}+\frac{\sqrt{3}}{3}$ b $\frac{2\sqrt{5}}{9}-\frac{\sqrt{5}}{6}$ c $\frac{-5\sqrt{10}}{6}+\frac{3\sqrt{10}}{8}$

Explanation	Solution
a Write the expression.	$\frac{\sqrt{3}}{2}+\frac{\sqrt{3}}{3}$
Multiply the denominators and cross multiply the numerators.	$=\frac{3\sqrt{3}+2\sqrt{3}}{6}$
Combine like surds.	$=\frac{5\sqrt{3}}{6}$
b Write the expression.	$\frac{2\sqrt{5}}{9}-\frac{\sqrt{5}}{6}$
Multiply the denominators and cross multiply the numerators.	$=\frac{12\sqrt{5}-9\sqrt{5}}{54}$
Combine like surds and simplify.	$=\frac{3\sqrt{5}}{54}=\frac{\sqrt{5}}{18}$
c Rewrite the expression.	$\frac{3\sqrt{10}}{8}-\frac{5\sqrt{10}}{6}$
Multiply the denominators and cross multiply the numerators.	$=\frac{18\sqrt{10}-40\sqrt{10}}{48}$
Combine like surds and simplify.	$=\frac{-22\sqrt{10}}{48}=-\frac{11\sqrt{10}}{24}$

Worked Example 5

Simplify each of the following. Assume x and y are positive real numbers.

a $\sqrt{xy} \times \sqrt{x^3y^5}$

b $\sqrt{12x^7y^2} \times \sqrt{6x^3y^4}$

Explanation	**Solution**
a Write the expression.	$\sqrt{xy} \times \sqrt{x^3y^5}$
Multiply what's inside the roots under one root.	$= \sqrt{xy \times x^3y^5}$
Multiply same bases and add the powers.	$= \sqrt{x^4y^6}$
Rewrite as a squared term.	$= \sqrt{(x^2x^3)^2}$
Simplify.	$= x^2y^3$
b Write the expression.	$\sqrt{12x^7y^2} \times \sqrt{6x^3y^4}$
Multiply what's inside the roots under one root.	$= \sqrt{12x^7y^2 \times 6x^3y^4}$
Combine terms.	$= \sqrt{72x^{10}y^6}$
Find the largest perfect square factor.	$= \sqrt{36x^{10}y^6 \times 2}$
Rewrite as a squared term multiplied by what's left.	$= \sqrt{(6x^5y^3)^2 \times 2}$
Split into separate square roots.	$= \sqrt{(6x^5y^3)^2} \times \sqrt{2}$
Simplify.	$= 6x^5y^3\sqrt{2}$

Your Turn To Try

1 **When simplified what is $\sqrt{5} \times \sqrt{2}$?**

A $\sqrt{7}$

B $\sqrt{10}$

C $2\sqrt{5}$

D $5\sqrt{2}$

2 **Simplify $2\sqrt{18} \times \sqrt{3}$. What is the result.**

A $6\sqrt{6}$

B $12\sqrt{3}$

C $4\sqrt{9}$

D $6\sqrt{2}$

3 **Which of these is the simplified version of $3\sqrt{5} \times 2\sqrt{10}$?**

A $30\sqrt{2}$

B $60\sqrt{5}$

C $6\sqrt{50}$

D $15\sqrt{10}$

4 **Simplify $2\sqrt{45} \times \sqrt{63}$. What is the result.**

A $18\sqrt{5}$

B $18\sqrt{7}$

C $18\sqrt{35}$

D $9\sqrt{35}$

5 **Find the simplified form of $\sqrt{12x^2} \times \sqrt{3}$, where x is a positive number?**

A $2x\sqrt{6}$

B $3x$

C $6x$

D $3x\sqrt{4}$

Question	1	2	3	4	5
Answer	B	A	A	C	C

Time To Practice

1 Simplify the following surds.

a $\sqrt{2} \times \sqrt{7}$ b $\sqrt{6} \times \sqrt{7}$ c $\sqrt{5} \times \sqrt{6}$

d $\sqrt{10} \times \sqrt{11}$ e $\sqrt{2} \times \sqrt{17}$ f $3\sqrt{2} \times \sqrt{21}$

g $2\sqrt{11} \times 5\sqrt{3}$ h $6\sqrt{3} \times 10\sqrt{7}$ i $3\sqrt{5} \times 4\sqrt{7}$

2 Simplify the following.

a $2\sqrt{8} \times \sqrt{2}$ b $3\sqrt{18} \times \sqrt{12}$ c $\sqrt{12} \times 4\sqrt{48}$

d $\sqrt{32} \times \sqrt{72}$ e $2\sqrt{75} \times 3\sqrt{45}$ f $\sqrt{20} \times \sqrt{125}$

g $4\sqrt{63} \times 2\sqrt{6}$ h $3\sqrt{50} \times 2\sqrt{20}$ i $4\sqrt{15} \times 2\sqrt{18}$

3 Simplify the following.

a $\frac{\sqrt{5}}{3} + \frac{\sqrt{5}}{4}$ b $\frac{\sqrt{2}}{5} - \frac{\sqrt{2}}{6}$ c $\frac{\sqrt{7}}{4} - \frac{\sqrt{7}}{12}$

d $\frac{2\sqrt{2}}{5} - \frac{\sqrt{2}}{2}$ e $\frac{3\sqrt{3}}{7} - \frac{\sqrt{3}}{2}$ f $\frac{7\sqrt{7}}{4} + \frac{\sqrt{7}}{12}$

g $\frac{\sqrt{8}}{3} - \frac{\sqrt{2}}{5}$ h $\frac{\sqrt{12}}{4} + \frac{\sqrt{3}}{6}$ i $\frac{2\sqrt{18}}{3} - \frac{\sqrt{72}}{2}$

4 Simplify the following surds, where x and y are positive numbers.

a $\sqrt{2x^2} \times \sqrt{3}$ b $\sqrt{8x^2y} \times \sqrt{y}$ c $\sqrt{2x^2y^2} \times 3\sqrt{27}$

d $\sqrt{12x^8y} \times 6\sqrt{2y}$ e $\sqrt{2x^6y^2} \times \sqrt{24x^4y^8}$ f $3\sqrt{7x^2y^3} \times 2\sqrt{x^2y^5}$

Challenge Yourself

1 Simplify the following surds, where x and y are positive numbers.

a $\frac{1}{2}\sqrt{60x^4} \times \sqrt{18x^2}$

b $\frac{1}{4}\sqrt{128x^8} \times 2\sqrt{75x^2}$

c $\frac{2}{3}\sqrt{12x^8y^2} \times 6\sqrt{32y^6}$

d $\frac{3}{8}\sqrt{63x^2y^2} \times \frac{6}{7}\sqrt{96y^{12}}$

e $\frac{2}{3}\sqrt{88x^4y^2} \times \frac{4}{5}\sqrt{48y^6}$

f $\frac{1}{6}\sqrt{200x^6y^4} \times \frac{3}{7}\sqrt{60x^8y^2}$

Dividing Surds: Keep 'Em Separate or Under One Roof (Root)

When dividing surds, there's just one rule.

But depending on the question, you can apply the same rule forward or backward.

Here's the general rule.

Surd Rules

Division

$$\frac{\sqrt{a}}{\sqrt{b}} = \sqrt{\frac{a}{b}}$$

$$\frac{m\sqrt{a}}{n\sqrt{b}} = \frac{m}{n}\sqrt{\frac{a}{b}}$$

The first way is going forward.

If there's a surd divided by another surd.

You can divide the numbers inside each root and write them under one root.

The second way is going backward.

If there's a fraction under a root.

You can split the fraction into two roots.

The square root of the top divided by the square root of the bottom.

OK, so when would you go forward, and when would you go backward?

You'd go forward if the numbers inside each root divide nicely into a whole number.

If they do.

Divide the numbers under one root and simplify.

If they don't.

Simplify the top and bottom surds separately.

As for going backward.

You'd do this if you're starting off with the root of a fraction.

For these, split it into two roots.

The square root of the top divided by the square root of the bottom.

Then simplify each separately.

Oh, and any numbers outside the roots?

Divide them but keep them outside the root.

What if there are letters,

and you don't know if dividing them will give a whole number?

Then what?

Do you go forward or backward?

Good question.

It really depends on what makes things easier.

Here's what to ask yourself.

Will putting them under one root make the powers easier to work with?

If yes.

Go forward and write them under one root.

If not.

Just keep them as separate roots and simplify each one.

There's no strict rule here.

It's more about what makes it easier to answer the question.

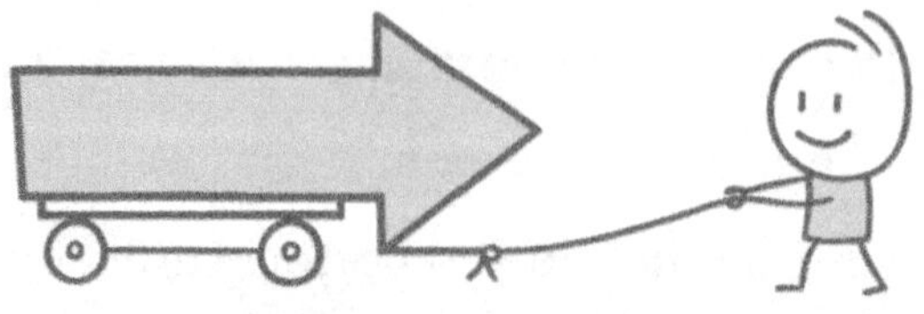

Worked Example 1

Evaluate the following.

a $\dfrac{\sqrt{15}}{\sqrt{3}}$ b $\dfrac{\sqrt{33}}{\sqrt{11}}$ c $\dfrac{\sqrt{42}}{\sqrt{7}}$

Explanation	**Solution**
a Write the expression.	$\dfrac{\sqrt{15}}{\sqrt{3}}$
Rewrite as a fraction under one square root.	$=\sqrt{\dfrac{15}{3}}$
Simplify the fraction.	$=\sqrt{5}$
b Write the expression.	$\dfrac{\sqrt{33}}{\sqrt{11}}$
Rewrite as a fraction under one square root.	$=\sqrt{\dfrac{33}{11}}$
Simplify the fraction.	$=\sqrt{3}$
c Write the expression.	$\dfrac{\sqrt{42}}{\sqrt{7}}$
Rewrite as a fraction under one square root.	$=\sqrt{\dfrac{42}{7}}$
Simplify the fraction.	$=\sqrt{6}$

Worked Example 2

Evaluate the following.

a $\dfrac{\sqrt{18}}{\sqrt{3}}$ b $\dfrac{\sqrt{8}}{\sqrt{2}}$ c $\dfrac{\sqrt{63}}{\sqrt{7}}$

Explanation	**Solution**
a Write the expression.	$\dfrac{\sqrt{18}}{\sqrt{3}}$
Rewrite as a fraction under one square root.	$= \sqrt{\dfrac{18}{3}}$
Simplify the fraction.	$= \sqrt{6}$
b Write the expression.	$\dfrac{\sqrt{8}}{\sqrt{2}}$
Rewrite as a fraction under one square root.	$= \sqrt{\dfrac{8}{2}}$
Simplify the fraction and square root.	$= \sqrt{4}$
	$= 2$
c Write the expression.	$\dfrac{\sqrt{63}}{\sqrt{7}}$
Rewrite as a fraction under one square root.	$= \sqrt{\dfrac{63}{7}}$
Simplify the fraction and square root.	$= \sqrt{9}$
	$= 3$

Worked Example 3

Evaluate the following.

a $\dfrac{\sqrt{60}}{2\sqrt{12}}$ b $\dfrac{3\sqrt{18}}{4\sqrt{6}}$

Explanation	**Solution**
a Write the expression.	$\dfrac{\sqrt{60}}{2\sqrt{12}}$
Rewrite as a fraction under one square root. Keep the number outside as is.	$=\dfrac{1}{2}\sqrt{\dfrac{60}{12}}$
Simplify the fraction inside the root.	$=\dfrac{1}{2}\sqrt{5}$
(Another - better - way to write it)	$=\dfrac{\sqrt{5}}{2}$
b Write the expression.	$\dfrac{3\sqrt{18}}{4\sqrt{6}}$
Rewrite as a fraction under one square root. Keep the number outside as is.	$=\dfrac{3}{4}\sqrt{\dfrac{18}{6}}$
Simplify the fraction inside the root.	$=\dfrac{3}{4}\sqrt{3}$
(Another - better - way to write it)	$=\dfrac{3\sqrt{3}}{4}$

Worked Example 4

Evaluate the following.

a $\dfrac{\sqrt{12}}{2} \times \dfrac{8}{\sqrt{3}}$

b $\dfrac{5\sqrt{8}}{2\sqrt{5}} \div \dfrac{\sqrt{2}}{\sqrt{10}}$

Explanation	**Solution**
a Write the expression.	$\dfrac{\sqrt{12}}{2} \times \dfrac{8}{\sqrt{3}}$
Multiply numerator and denominators.	$= \dfrac{8\sqrt{12}}{2\sqrt{3}}$
Rewrite as a fraction under one square root. Simplify the numbers outside.	$= 4\sqrt{\dfrac{12}{3}}$
Simplify the fraction and the square root.	$= 4\sqrt{4} = 4(2)$
	$= 8$
b Write the expression.	$\dfrac{5\sqrt{8}}{2\sqrt{5}} \div \dfrac{\sqrt{2}}{\sqrt{10}}$
Flip the second fraction and change sign to a multiplication.	$= \dfrac{5\sqrt{8}}{2\sqrt{5}} \times \dfrac{\sqrt{10}}{\sqrt{2}}$
Multiply fractions and rewrite as a single fraction under one square root.	$= \dfrac{5\sqrt{80}}{2\sqrt{10}} = \dfrac{5}{2}\sqrt{\dfrac{80}{10}}$
Simplify the fraction and the square root. Keep the number outside as is.	$= \dfrac{5}{2}\sqrt{8} = \dfrac{5}{2} \times 2\sqrt{2}$
Simplify.	$= 5\sqrt{2}$

Worked Example 5

Evaluate the following.

a $\sqrt{\dfrac{1}{16}}$ b $\sqrt{\dfrac{8}{9}}$ c $2\sqrt{\dfrac{18}{25}}$

Explanation	**Solution**
a Write the expression.	$\sqrt{\dfrac{1}{16}}$
Split into two separate roots and simplify.	$=\dfrac{\sqrt{1}}{\sqrt{16}}=\dfrac{1}{4}$
b Write the expression.	$\sqrt{\dfrac{8}{9}}$
Split into two separate roots and simplify each root.	$=\dfrac{\sqrt{8}}{\sqrt{9}}=\dfrac{\sqrt{4\times 2}}{3}$
	$=\dfrac{2\sqrt{2}}{3}$
c Write the expression.	$2\sqrt{\dfrac{18}{25}}$
Split into two separate roots and simplify each root.	$=\dfrac{2\sqrt{18}}{\sqrt{25}}=\dfrac{2\sqrt{9\times 2}}{5}$
Simplify fraction.	$=\dfrac{2\times 3\sqrt{2}}{5}=\dfrac{6\sqrt{2}}{5}$

Worked Example 6

Simplify each of the following. Assume x and y are positive real numbers.

a $\dfrac{\sqrt{x^4y^3}}{\sqrt{x^2y}}$

b $\dfrac{\sqrt{16x^7y^9}}{\sqrt{8xy}}$

Explanation	Solution
a Write the expression.	$\dfrac{\sqrt{x^4y^3}}{\sqrt{x^2y}}$
Rewrite as a fraction under one square root.	$= \sqrt{\dfrac{x^4y^3}{x^2y}}$
Simplify the fraction and write as a perfect square.	$= \sqrt{x^2y^2} = \sqrt{(xy)^2}$
Simplify the square root.	$= xy$
b Write the expression.	$\dfrac{\sqrt{16x^7y^9}}{\sqrt{8xy}}$
Rewrite as a fraction under one square root.	$= \sqrt{\dfrac{16x^7y^9}{8xy}}$
Simplify the fraction.	$= \sqrt{2x^6y^8}$
Rewrite a s perfect square factor multiplied by what's left.	$= \sqrt{2 \times (x^3y^4)^2}$
Simplify the square root.	$= x^3y^4\sqrt{2}$

$\sqrt{a} \div \sqrt{b}$ is the same as $\dfrac{\sqrt{a}}{\sqrt{b}}$.

Worked Example 7

Simplify the following. Assume a and b are positive real numbers. $\dfrac{\sqrt{a^3 b}}{3\sqrt{a^9 b^7}}$

Explanation	Solution
Write the expression.	$\dfrac{\sqrt{a^3b}}{3\sqrt{a^9b^7}}$
Rewrite as a fraction under one square root.	$=\dfrac{1}{3}\sqrt{\dfrac{a^3b}{a^9b^7}}$
Simplify the fraction using Exponent Law 2.	$=\dfrac{1}{3}\sqrt{\dfrac{1}{a^6b^6}}$
Rewrite denominator as a perfect square.	$=\dfrac{1}{3}\sqrt{\dfrac{1}{(a^3b^3)^2}}$
Split the square root into two roots being divided.	$=\dfrac{1}{3}\times\dfrac{\sqrt{1}}{\sqrt{(a^3b^3)^2}}$
Simplify the denominator.	$=\dfrac{1}{3}\times\dfrac{1}{a^3b^3}$
	$=\dfrac{1}{3a^3b^3}$

> Forgetting that the square root of a fraction can be split into two square roots. The square root of the top divided by the square root of the bottom. ⚠

Your Turn To Try

1 **What is the simplified result of $\sqrt{48} \div \sqrt{12}$?**

A 4

B 6

C 2

D 3

2 **Simplify $\sqrt{18} \div \sqrt{2}$. What is the result.**

A 3

B 8

C 4

D 9

3 **What is $\sqrt{75} \div \sqrt{5}$?**

A 5

B $\sqrt{15}$

C 3

D $\sqrt{10}$

4 **Choose the simplified version of $3\sqrt{27} \div \sqrt{3}$. What is the result.**

A 6

B $3\sqrt{3}$

C 9

D $9\sqrt{3}$

5 **Find the simplified form of $\sqrt{72x^2} \div \sqrt{8}$, where x is a positive number?**

A 3

B $3\sqrt{2x}$

C $9x$

D $3x$

Question	1	2	3	4	5
Answer	C	A	B	C	D

Time To Practice

1 Simplify the following.

a $\sqrt{16} \div \sqrt{4}$

b $\sqrt{25} \div \sqrt{5}$

c $\sqrt{81} \div \sqrt{9}$

d $\sqrt{64} \div \sqrt{2}$

e $3\sqrt{36} \div 2\sqrt{4}$

f $2\sqrt{45} \div \sqrt{5}$

g $4\sqrt{72} \div 2\sqrt{18}$

h $\sqrt{200} \div 2\sqrt{5}$

i $8\sqrt{18} \div 4\sqrt{2}$

j $\sqrt{\frac{9}{64}}$

k $\sqrt{\frac{12}{25}}$

l $3\sqrt{\frac{27}{49}}$

2 Simplify the following surds.

a $\frac{3\sqrt{64}}{5\sqrt{8}}$

b $\frac{\sqrt{60}}{5\sqrt{5}}$

c $\frac{9\sqrt{80}}{12\sqrt{4}}$

d $\frac{3\sqrt{88}}{14\sqrt{11}}$

e $\frac{3\sqrt{108}}{11\sqrt{6}}$

f $\frac{12\sqrt{56}}{18\sqrt{2}}$

3 Simplify and write the answers in the simplest surd form.

a $\frac{1}{\sqrt{2}} \times \frac{\sqrt{6}}{3}$

b $\frac{\sqrt{10}}{\sqrt{3}} \times \frac{\sqrt{6}}{2}$

c $\frac{\sqrt{12}}{3} \times \frac{\sqrt{9}}{2\sqrt{6}}$

d $\frac{\sqrt{5}}{2} \div \frac{4}{\sqrt{10}}$

e $\frac{\sqrt{18}}{\sqrt{2}} \div \frac{3}{\sqrt{6}}$

f $\frac{5}{\sqrt{6}} \div \frac{2}{9\sqrt{48}}$

4 Simplify and write the answers in the simplest form.

a $\frac{\sqrt{x^8y^4}}{\sqrt{x^2y^2}}$

b $\frac{\sqrt{8x^8y^4}}{\sqrt{2x^4y^6}}$

c $\frac{\sqrt{9x^2}}{\sqrt{x^4y^8}}$

d $\frac{\sqrt{x^4y^2}}{\sqrt{x^6y^4}} \div \frac{\sqrt{x^8}}{\sqrt{y^6}}$

e $\frac{\sqrt{xy}}{\sqrt{x^5y^7}} \div \frac{\sqrt{x^2y^4}}{\sqrt{12x^8y^{12}}}$

f $\frac{2\sqrt{6x^5y^3}}{\sqrt{x^3y}} \div \frac{3\sqrt{y}}{\sqrt{y^7}}$

Challenge Yourself

1 Simplify the following surds.

a $\dfrac{9\sqrt{12} \times 2\sqrt{18}}{3\sqrt{8}}$

b $\left(\dfrac{(5\sqrt{3})^2 \times 2\sqrt{5}}{3}\right) \times \dfrac{\sqrt{3}}{6\sqrt{5}}$

c $\dfrac{16}{\sqrt{3}} \div \left(\dfrac{(2\sqrt{2})^2 \times (2\sqrt{5})^2}{5\sqrt{60}}\right)$

d $\left(\dfrac{\sqrt{54} \times 3\sqrt{22}}{3\sqrt{8}}\right) \div \dfrac{16\sqrt{33}}{\sqrt{3}}$

2 Simplify the following surds, where x and y are positive numbers.

a $\dfrac{\sqrt{x^8y^{12}}}{\sqrt{3x^2y^3}} \div \dfrac{4\sqrt{x^5y^7}}{\sqrt{21x}}$

b $\dfrac{2\sqrt{5x^4y^2}}{\sqrt{4x^3y^5}} \div \dfrac{3\sqrt{10x^7y}}{\sqrt{6y^6}}$

Rationalizing the Denominator: Can't Be Having Surds in The Bottom Now

So far, when dividing surds, the denominator ended up being a whole number.

But that doesn't happen all the time.

Sometimes the denominator ends up being a surd.

Now.

In math, we don't like having surds in the denominator.

It's fine in the numerator.

But the denominator?

Nah, we only want whole numbers there.

So, how do we make the denominator a whole number?

There's a process for that.

It's called *Rationalizing the Denominator*.

Here's the rule for it.

Surd Rules

Rationalizing the Denominator

$$\frac{a}{\sqrt{b}} = \frac{a}{\sqrt{b}} \times \frac{\sqrt{b}}{\sqrt{b}}$$

$$= \frac{a\sqrt{b}}{b}$$

If there's a surd in the denominator,

multiply the top and bottom of the fraction by that same surd in the denominator.

To show we're doing this in one step we write it as a separate fraction.

It just keeps things neat.

Multiplying across gives you an equivalent fraction.

Where the denominator is a whole number.

Which is what we want.

From there, as always, simplify if you can.

That's the process in a nutshell.

But why do we do this?

Isn't it just creating more work?

Yeah, it is.

It comes down to being as accurate as possible.

If the denominator is a surd, you're dividing by a number that has endless decimals.

How do you divide by a number that never ends?

Short answer.

You don't.

It's not practical, or it'd be a real mess.

Dividing by a whole number keeps things neat and accurate.

Here's a tip before you give it a shot.

Always simplify any surds in the denominator first.

Before multiplying the top and the bottom of the fraction.

Get the number inside the root as small as possible.

It makes the maths much easier.

So, to sum it up, here's what to do.

1) Simplify the surd in the denominator (if you can)
2) Multiply the top and bottom of the fraction by that same (simplified) surd
3) Simplify further if needed

Worked Example 1

Rationalize the denominator for the following.

a $\frac{2}{\sqrt{3}}$ b $\frac{4}{\sqrt{5}}$

Explanation | **Solution**

a Write the expression.

$$\frac{2}{\sqrt{3}}$$

Multiply both top and bottom by the surd in the denominator.

$$= \frac{2}{\sqrt{3}} \times \frac{\sqrt{3}}{\sqrt{3}}$$

Combine and simplify the expression.

$$= \frac{2\sqrt{3}}{\left(\sqrt{3}\right)^2}$$

$$= \frac{2\sqrt{3}}{3}$$

b Write the expression.

$$\frac{4}{\sqrt{5}}$$

Multiply both top and bottom by the surd in the denominator.

$$= \frac{4}{\sqrt{5}} \times \frac{\sqrt{5}}{\sqrt{5}}$$

Combine and simplify the expression.

$$= \frac{4\sqrt{5}}{\left(\sqrt{5}\right)^2}$$

$$= \frac{4\sqrt{5}}{5}$$

When squaring a surd, no need to write $\left(\sqrt{a}\right)^2$ in your steps. Just write a.

Worked Example 2

Express the following in their simplest form with a rational denominator.

a $\frac{\sqrt{2}}{\sqrt{7}}$ b $\frac{\sqrt{8}}{\sqrt{3}}$

Explanation		**Solution**
a	Write the expression.	$\frac{\sqrt{2}}{\sqrt{7}}$
	Multiply both top and bottom by the surd in the denominator.	$= \frac{\sqrt{2}}{\sqrt{7}} \times \frac{\sqrt{7}}{\sqrt{7}}$
	Simplify the expression.	$= \frac{\sqrt{14}}{7}$
b	Write the expression.	$\frac{\sqrt{8}}{\sqrt{3}}$
	Multiply both top and bottom by the surd in the denominator.	$= \frac{\sqrt{8}}{\sqrt{3}} \times \frac{\sqrt{3}}{\sqrt{3}}$
	Find the largest perfect square factor.	$= \frac{\sqrt{24}}{3}$
	Take out the perfect square factor.	$= \frac{\sqrt{4 \times 6}}{3}$ $= \frac{2\sqrt{6}}{3}$

After rationalizing the denominator, always double-check if you can simplify the resulting fraction.

Worked Example 3

Express the following in their simplest form with a rational denominator.

a $\frac{7\sqrt{3}}{\sqrt{10}}$

b $\frac{5\sqrt{11}}{6\sqrt{5}}$

Explanation	Solution
a Write the expression.	$\frac{7\sqrt{3}}{\sqrt{10}}$
Multiply both top and bottom by the surd in the denominator.	$= \frac{7\sqrt{3}}{\sqrt{10}} \times \frac{\sqrt{10}}{\sqrt{10}}$
Combine and simplify the expression.	$= \frac{7\sqrt{30}}{10}$
b Write the expression.	$\frac{5\sqrt{11}}{6\sqrt{5}}$
Multiply both top and bottom by the surd in the denominator.	$= \frac{5\sqrt{11}}{6\sqrt{5}} \times \frac{\sqrt{5}}{\sqrt{5}}$
Combine and simplify the expression.	$= \frac{5\sqrt{55}}{6 \times 5}$
Simplify the fraction by cancelling the common factor.	$= \frac{5\sqrt{55}}{30}$
	$= \frac{\sqrt{55}}{6}$

Worked Example 4

Rationalize the denominator. $\frac{9\sqrt{6}}{2\sqrt{3}}$

Explanation	Solution
Write the expression.	$\frac{9\sqrt{6}}{2\sqrt{3}}$
Multiply both top and bottom by the surd in the denominator.	$= \frac{9\sqrt{6}}{2\sqrt{3}} \times \frac{\sqrt{3}}{\sqrt{3}}$
Find the largest perfect square factor.	$= \frac{9\sqrt{18}}{2 \times 3}$
Take out the perfect square factor.	$= \frac{9\sqrt{9 \times 2}}{2 \times 3}$
Simplify the fraction by cancelling the common factor.	$= \frac{9 \times 3\sqrt{2}}{2 \times 3}$
	$= \frac{9\sqrt{2}}{2}$

Thinking you can't have a surd in the **numerator** and trying to 'rationalize' by multiplying the top and bottom by the surd in the **numerator**.

Worked Example 5

Rationalize the denominator. $\dfrac{5\sqrt{8}}{3\sqrt{27}}$

Explanation	**Solution**
Write the expression.	$\dfrac{5\sqrt{8}}{3\sqrt{27}}$
Find the largest perfect square factor for both surds.	$=\dfrac{5\sqrt{4\times 2}}{3\sqrt{9\times 3}}$
Take out the perfect square factors.	$=\dfrac{5\times 2\sqrt{2}}{3\times 3\sqrt{3}}$
Multiply the numbers outside the square root.	$=\dfrac{10\sqrt{2}}{9\sqrt{3}}$
Multiply both top and bottom by the surd in the denominator.	$=\dfrac{10\sqrt{2}}{9\sqrt{3}}\times\dfrac{\sqrt{3}}{\sqrt{3}}$
Simplify.	$=\dfrac{10\sqrt{6}}{9\times 3}$
	$=\dfrac{10\sqrt{6}}{27}$

Your Turn To Try

1 **Which of the following is the correct rationalization of $\frac{1}{\sqrt{2}}$?**

A $2\sqrt{2}$

B $\sqrt{2}$

C $\frac{1}{2}$

D $\frac{\sqrt{2}}{2}$

2 **Choose the correct rationalized form of $\frac{4}{\sqrt{3}}$.**

A $\frac{\sqrt{3}}{4}$

B $\frac{4\sqrt{3}}{3}$

C 4

D $4\sqrt{3}$

3 **For $\frac{3}{\sqrt{12}}$, what should you multiply top and bottom by to rationalize?**

A $\sqrt{4}$

B $\sqrt{3}$

C $\sqrt{6}$

D $\sqrt{2}$

4 **What is the simplified form of $\frac{10}{\sqrt{18}}$ after rationalizing the denominator?**

A $\frac{5\sqrt{2}}{3}$

B $\frac{5\sqrt{3}}{9}$

C $\frac{2\sqrt{3}}{5}$

D $\frac{3\sqrt{2}}{2}$

5 **Which expression is equal to $\frac{\sqrt{20}}{\sqrt{50}}$?**

A $\frac{5\sqrt{2}}{2}$

B $\frac{4\sqrt{5}}{5}$

C $\frac{\sqrt{10}}{5}$

D $\frac{2\sqrt{5}}{5}$

Question	1	2	3	4	5
Answer	D	B	B	A	C

Time To Practice

1 Rationalize the denominators.

a $\frac{1}{\sqrt{7}}$ b $\frac{4}{\sqrt{5}}$ c $\frac{8}{\sqrt{2}}$

d $\frac{\sqrt{5}}{\sqrt{3}}$ e $\frac{\sqrt{7}}{\sqrt{2}}$ f $\frac{\sqrt{11}}{\sqrt{2}}$

2 Rationalize the denominators.

a $\frac{4\sqrt{2}}{\sqrt{7}}$ b $\frac{5\sqrt{2}}{\sqrt{3}}$ c $\frac{3\sqrt{6}}{\sqrt{7}}$

d $\frac{2\sqrt{7}}{5\sqrt{10}}$ e $\frac{4\sqrt{7}}{5\sqrt{3}}$ f $\frac{4\sqrt{3}}{3\sqrt{5}}$

g $\frac{8\sqrt{3}}{7\sqrt{7}}$ h $\frac{9\sqrt{6}}{2\sqrt{3}}$ i $\frac{5\sqrt{12}}{9\sqrt{6}}$

3 Express the following in their simplest form with a rational denominator.

a $\frac{4\sqrt{10}}{\sqrt{18}}$ b $\frac{8\sqrt{20}}{\sqrt{28}}$ c $\frac{2\sqrt{45}}{5\sqrt{12}}$

d $\frac{5\sqrt{50}}{3\sqrt{75}}$ e $\frac{6\sqrt{44}}{2\sqrt{20}}$ f $\frac{5\sqrt{63}}{8\sqrt{32}}$

g $\frac{3\sqrt{128}}{4\sqrt{45}}$ h $\frac{3\sqrt{125}}{5\sqrt{48}}$ i $\frac{2\sqrt{98}}{7\sqrt{27}}$

Challenge Yourself

1 Rationalize the denominator and simplify.

a $\dfrac{3\sqrt{3}}{2} \times \dfrac{5}{\sqrt{6}}$

b $\dfrac{5\sqrt{18}}{3} \times \dfrac{\sqrt{5}}{6\sqrt{3}}$

c $\dfrac{\sqrt{20}}{5\sqrt{2}} \times \dfrac{2}{\sqrt{27}}$

d $\dfrac{\sqrt{12}}{3\sqrt{5}} \times \dfrac{4\sqrt{15}}{\sqrt{2}}$

2 Simplify the following with a rational denominator.

a $\dfrac{3\sqrt{28}}{\sqrt{18}} \div \dfrac{9\sqrt{12}}{2\sqrt{4}}$

b $\dfrac{4\sqrt{3}}{\sqrt{72}} \div \dfrac{3\sqrt{5}}{2\sqrt{12}}$

3 Show that the following is a whole number and state this number.

a $\dfrac{\sqrt{125} - \sqrt{80}}{\sqrt{98}}$

b $\dfrac{\sqrt{75} - \sqrt{27}}{\sqrt{90}}$

3. Answers

Identifying Surds

Time to Practice

1	a	Yes	b	No	c	Yes
	d	Yes	e	Yes	f	No
	g	Yes	h	No	i	Yes
	j	No	k	Yes	l	No
	m	No	n	No	o	Yes
	p	No	q	Yes	r	Yes

Challenge Yourself

1	a	No	b	Yes	c	No
	d	No	e	Yes	f	No

2 $\sqrt{\frac{1}{8}}\ and \sqrt{21}$

Simplifying Surds

Time to Practice

1 a $2\sqrt{3}$ b $3\sqrt{2}$ c $2\sqrt{5}$

d $2\sqrt{6}$ e $3\sqrt{3}$ f $6\sqrt{2}$

g $7\sqrt{2}$ h $5\sqrt{5}$ i $6\sqrt{10}$

2 a $6\sqrt{2}$ b $6\sqrt{5}$ c $6\sqrt{7}$

d $16\sqrt{3}$ e $35\sqrt{2}$ f $40\sqrt{2}$

g $20\sqrt{6}$ h $21\sqrt{3}$ i $30\sqrt{3}$

3 a $\sqrt{5}$ b $\frac{2\sqrt{7}}{3}$ c $\frac{3\sqrt{6}}{4}$

d $\frac{\sqrt{6}}{3}$ e $\frac{4\sqrt{5}}{5}$ f $\frac{\sqrt{11}}{3}$

g $\frac{\sqrt{21}}{4}$ h $\frac{2\sqrt{14}}{7}$ i $\frac{2\sqrt{17}}{3}$

4 a $\sqrt{50}$ b $\sqrt{27}$ c $\sqrt{128}$

d $\sqrt{700}$ e $\sqrt{810}$ f $\sqrt{99}$

Challenge Yourself

1 3 2 15 3 $a = 20, b = 3$

4 a $2\sqrt{2}$ b $\sqrt{2}$

c $\frac{3\sqrt{7}}{2}$ d $\frac{3\sqrt{3}}{2}$

Square and Square Roots of Surds

Time to Practice

1 a 7 b 11 c 13
d $5x$ e $9x$ f $3x^2$
g $8x$ h $48x$ i $50x^2$

2 a z b $5z$ c $11z$
d $6b$ e $10b$ f $7b$
g $12pq$ h $15p^3q^2$ i $17pq^4$

3 a $2x\sqrt{3}$ b $2x\sqrt{7}$ c $3y\sqrt{5}$
d $4x^4\sqrt{3}$ e $5x^3\sqrt{2}$ f $6y^2\sqrt{2}$
g $5x^4y^3\sqrt{3}$ h $4x^2y\sqrt{6}$ i $7xy^3\sqrt{2}$

4 a $a^6\sqrt{a}$ b $b^2\sqrt{2b}$ c $2a^3\sqrt{a}$
d $12a^4\sqrt{a}$ e $4a\sqrt{ab}$ f $3a^3b^2\sqrt{5ab}$
g $3b^2\sqrt{14ab}$ h $12a^7b\sqrt{2ab}$ i $5a^2b^4\sqrt{15ab}$

Challenge Yourself

1 a $15x^2y^2\sqrt{5}$ b $63xy^3\sqrt{2}$ c $x^2y^6\sqrt{22}$
d $\frac{12x^9y^8\sqrt{5}}{5}$ e $\frac{x^2y^3z^4\sqrt{3}}{3}$ f $2x^4yz^2\sqrt{6}$
g $6x^4yz^2\sqrt{3xyz}$ h $15x^3z^2\sqrt{3xyz}$

Adding and Subtracting Surds

Time to Practice

1 a $7\sqrt{7}$ b $13\sqrt{3}$ c $15\sqrt{2}$
d $4\sqrt{3}$ e $5\sqrt{10}$ f $14\sqrt{5}$
g $\sqrt{13}$ h $2\sqrt{21}$

2 a $15\sqrt{5}+5\sqrt{3}$ b $17\sqrt{3}-18\sqrt{7}$
c $-3\sqrt{2}-3\sqrt{10}$ d $3\sqrt{11}+2\sqrt{15}$

3 a $5\sqrt{2}$ b $\sqrt{5}$ c $10\sqrt{3}$
d $7\sqrt{6}$ e $32\sqrt{2}$ f $\sqrt{5}$
g $5\sqrt{5}+6\sqrt{7}$ h $6-9\sqrt{3}$

4 a $8\sqrt{x}+3\sqrt{y}$ b $13\sqrt{x}-4\sqrt{y}$
c $6\sqrt{y}+\sqrt{x}$ d $-\sqrt{x}-34\sqrt{y}$

Challenge Yourself

1 $a=8, b=2$ 2 $a=9, b=5$ 3 $a=11, b=2$

4 a $-19\sqrt{3a}$ b $14\sqrt{2b}-12$
c $-5\sqrt{2a}$ d $\dfrac{10\sqrt{6ab}}{3}$

Multiplying Surds

Time to Practice

1 a $\sqrt{14}$ b $\sqrt{42}$ c $\sqrt{30}$
d $\sqrt{110}$ e $\sqrt{34}$ f $3\sqrt{42}$
g $10\sqrt{33}$ h $60\sqrt{21}$ i $12\sqrt{35}$

2 a 8 b $18\sqrt{6}$ c 96
d 48 e $90\sqrt{15}$ f 50
g $24\sqrt{42}$ h $60\sqrt{10}$ i $24\sqrt{30}$

3 a $\frac{7\sqrt{5}}{12}$ b $\frac{\sqrt{2}}{30}$ c $\frac{\sqrt{7}}{6}$
d $-\frac{\sqrt{2}}{10}$ e $-\frac{\sqrt{3}}{14}$ f $\frac{11\sqrt{7}}{6}$
g $\frac{7\sqrt{2}}{15}$ h $\frac{2\sqrt{3}}{3}$ i $-\sqrt{2}$

4 a $x\sqrt{6}$ b $2xy\sqrt{2}$ c $9xy\sqrt{6}$
d $12x^4y\sqrt{6}$ e $4x^5y^5\sqrt{3}$ f $6x^2y^4\sqrt{7}$

Challenge Yourself

4 a $3x^3\sqrt{30}$ b $20x^5\sqrt{6}$ c $32x^4y^4\sqrt{6}$
d $\frac{27xy^7\sqrt{42}}{7}$ e $\frac{64x^2y^4\sqrt{66}}{15}$ f $\frac{10x^7y^3\sqrt{30}}{7}$

Dividing Surds

Time to Practice

1 a 2 b $\sqrt{5}$ c 3

d $4\sqrt{2}$ e $\frac{9}{2}$ f 6

g 4 h $\sqrt{10}$ i 6

j $\frac{3}{8}$ k $\frac{2\sqrt{3}}{5}$ l $\frac{9\sqrt{3}}{7}$

2 a $\frac{6\sqrt{2}}{5}$ b $\frac{2\sqrt{3}}{5}$ c $\frac{3\sqrt{5}}{2}$

d $\frac{3\sqrt{2}}{7}$ e $\frac{9\sqrt{2}}{11}$ f $\frac{4\sqrt{7}}{3}$

3 a $\frac{\sqrt{3}}{3}$ b $\sqrt{5}$ c $\frac{\sqrt{2}}{2}$

d $\frac{5\sqrt{2}}{8}$ e $\sqrt{6}$ f $45\sqrt{2}$

4 a x^3y b $\frac{2x^2}{y}$ c $\frac{3}{xy^4}$

d $\frac{y^2}{x^5}$ e $2xy\sqrt{3}$ f $\frac{2xy^4\sqrt{6}}{3}$

Challenge Yourself

1 a $18\sqrt{3}$ b $\frac{25\sqrt{3}}{3}$

c $\sqrt{5}$ d $\frac{3\sqrt{6}}{32}$

2 a $\frac{xy\sqrt{7}}{4}$ b $\frac{y\sqrt{3}}{3x^3}$

Rationalizing the Denominator

Time to Practice

1 a $\frac{\sqrt{7}}{7}$ b $\frac{4\sqrt{5}}{5}$ c $4\sqrt{2}$

d $\frac{\sqrt{15}}{3}$ e $\frac{\sqrt{14}}{2}$ f $\frac{\sqrt{22}}{2}$

2 a $\frac{4\sqrt{14}}{7}$ b $\frac{5\sqrt{6}}{3}$ c $\frac{3\sqrt{42}}{7}$

d $\frac{\sqrt{70}}{25}$ e $\frac{4\sqrt{21}}{15}$ f $\frac{4\sqrt{15}}{15}$

g $\frac{8\sqrt{21}}{49}$ h $\frac{9\sqrt{2}}{2}$ i $\frac{5\sqrt{2}}{9}$

3 a $\frac{4\sqrt{5}}{3}$ b $\frac{8\sqrt{35}}{7}$ c $\frac{\sqrt{15}}{5}$

d $\frac{5\sqrt{6}}{9}$ e $\frac{3\sqrt{55}}{5}$ f $\frac{15\sqrt{14}}{64}$

g $\frac{2\sqrt{10}}{5}$ h $\frac{\sqrt{15}}{4}$ i $\frac{2\sqrt{6}}{9}$

Challenge Yourself

1 a $\frac{15\sqrt{2}}{4}$ b $\frac{5\sqrt{30}}{18}$

c $\frac{2\sqrt{30}}{45}$ d $4\sqrt{2}$

2 a $\frac{2\sqrt{42}}{27}$ b $\frac{4\sqrt{10}}{15}$

3 a $\frac{\sqrt{10}}{14}$ b $\frac{\sqrt{30}}{15}$

Linear
Equations

Basic Equations: Moving Things Around To Find x or y

Everything so far has been about simplifying stuff.

Like fractions.

Exponents.

Surds.

Skills you'll use over and over again.

Now, we'll look at solving linear equations.

What's a linear equation?

An equation is just two expressions with an equal sign in between.

What's on the left equals what's on the right.

A linear equation has a letter and numbers on either side of the equals sign.

The letter can be anything, x,y, z, p, t, whatever.

What makes it linear is that the letter has a power of 1.

It can't have any other power.

No squares.

No cubes.

No fractions

No negatives.

Nothing.

Just the letter.

It can be multiplied by a number.

But its power has to be 1.

This letter is called the **variable**.

Solving an equation means working out what the variable equals.

What number can replace the variable so both sides of the equation will be the same.

To find this number, we move things around.
In a way where the variable ends up on its own on one side.
And the numbers are on the other.

This is called "making x the subject" or "solving for x".
Or whatever the variable is.

Here's how to do it.
In any equation, there are two kinds of numbers.

- **Coefficient** – the number in front of the variable.
- **Constant** – the number without a variable.

Sometimes there's one constant.
Sometimes there are two — one constant on each side of the equals sign.

To get the variable by itself,
start by moving the constants to one side.
The side that doesn't have the variable.
Then you move the coefficient to the other side too.
This will leave the variable by itself.

Whenever you move a number across the equals sign,
you do what's called the *opposite operation.*

An operation is basically what you do to a number.
Add, subtract, multiply, or divide.
And the opposite operation just means that.
You do the opposite.

For example.

If a constant is added, subtract it on the other side.

If it's subtracted, then add it.

Same with the coefficient.

If it's multiplied, divide it on the other side.

If it's divided, then multiply it.

Now, here's the important bit.

When you move a constant across the equals sign, its sign changes.

But when you move a coefficient, its sign doesn't change.

Just a quick tip.

Any time you move a number, left or right, simplify as you go.

It keeps things neat.

And helps avoid mistakes.

Oh, and one more thing.

When multiplying or dividing,

don't forget how the sign rules work.

When the signs are the same.

You get a positive.

When they're different.

You get a negative.

Sign Rules

$+$	$\times$ *or* $\div$	$+$	$=$	$+$
$-$	$\times$ *or* $\div$	$-$	$=$	$+$
$+$	$\times$ *or* $\div$	$-$	$=$	$-$
$-$	$\times$ *or* $\div$	$+$	$=$	$-$

Alright, 3 steps to solve basic linear equations.

1) Move the constants to the opposite side of the variable.
2) Move the coefficient to the other side.
3) Simplify if needed.

Worked Example 1

Solve the following linear equations.

a $x - 4 = 6$ b $x + 2 = 15$ c $1 + y = -5$

Explanation		**Solution**
a	Write the equation.	$x - 4 = 6$
	Move the constant to the other side.	$x = 6 + 4$
	Simplify.	$x = 10$
b	Write the equation.	$x + 2 = 15$
	Move the constant to the other side.	$x = 15 - 2$
	Simplify.	$x = 13$
c	Write the equation.	$1 + y = -5$
	Move the constant to the other side.	$y = -5 - 1$
	Simplify.	$y = -6$

Equations that require one step to solve are called one-step equations.
Equations that require two step to solve are called two-step equations.

Worked Example 2

Solve the following linear equations.

a $\mathbf{5x = 35}$ b $\boldsymbol{\frac{x}{4} = 3}$ c $\boldsymbol{-3a = 15}$

Explanation		**Solution**
a	Write the equation.	$5x = 35$
	Move the coefficient to the other side.	$x = \frac{35}{5}$
	Simplify.	$x = 7$
b	Write the equation.	$\frac{x}{4} = 3$
	Move the coefficient to the other side.	$x = 3 \times 4$
	Simplify.	$x = 12$
c	Write the equation.	$-3a = 15$
	Move the coefficient to the other side.	$a = -\frac{15}{3}$
	Simplify.	$a = -5$

To move a fraction coefficient $\frac{a}{b}$, multiply the other side of the equal sign by its reciprocal $\frac{b}{a}$.

Worked Example 3

Solve the following linear equations.

a $\mathbf{2x - 4 = 6}$ b $\mathbf{3x + 2 = 20}$ c $\mathbf{1 - 2y = -5}$

	Explanation	**Solution**
a	Write the equation.	$2x - 4 = 6$
	Move the constant to the other side.	$2x = 6 + 4$
	Simplify right hand side.	$2x = 10$
	Move the coefficient to the other side.	$x = \frac{10}{2}$
	Simplify.	$x = 5$
b	Write the equation.	$3x + 2 = 20$
	Move the constant to the other side.	$3x = 20 - 2$
	Simplify right hand side.	$3x = 18$
	Move the coefficient to the other side.	$x = \frac{18}{3}$
	Simplify.	$x = 6$
c	Write the equation.	$1 - 2y = -5$
	Move the constant to the other side.	$-2y = -5 - 1$
	Simplify right hand side.	$-2y = -6$
	Move the coefficient to the other side.	$y = \frac{-6}{-2}$
	Simplify.	$y = 3$

Worked Example 4

Solve the following linear equations.

a $\frac{1}{2}x+3=5$

b $\frac{7}{4}+\frac{3}{2}y=5$

Explanation		**Solution**
a	Write the equation.	$\frac{1}{2}x+3=5$
	Move the constant to the other side.	$\frac{1}{2}x=2$
	Move the coefficient to the other side.	$x=2\times 2$
	Simplify.	$x=4$
b	Write the equation.	$\frac{7}{4}+\frac{3}{2}y=5$
	Move the constant to the other side.	$\frac{3}{2}y=5-\frac{7}{4}$
	Simplify the right side.	$\frac{3}{2}y=\frac{20}{4}-\frac{7}{4}$
		$\frac{3}{2}y=\frac{13}{4}$
	Move the coefficient to the other side.	$y=\frac{13}{4}\times\frac{2}{3}$
	Simplify.	$y=\frac{13}{6}$

Your Turn To Try

1 **What does x equal in this equation $7x = 28$?**

A 3 B 7

C 5 D 4

2 **Which of the following is the correct value of x for $\frac{5x}{3} = 10$?**

A 2 B 10

C 6 D 9

3 **What the value of x in the equation $3x + 5 = 11$?**

A 5 B 2

C 3 D 4

4 **Find the correct value of x for $\frac{x}{6} - 1 = 2$?**

A 12 B 16

C 14 D 18

5 **What value of x makes this equation $\frac{3x}{4} + \frac{1}{2} = \frac{7}{2}$ true?**

A 4 B 3

C 2 D 1

Question	1	2	3	4	5
Answer	D	C	B	D	A

Time To Practice

1 Solve the following equations.

a $x + 6 = 8$ b $x - 7 = 4$ c $8 - x = 3$

d $6a = -42$ e $9a = 63$ f $4a = -44$

g $\frac{m}{4} = 3$ h $\frac{3m}{5} = 12$ i $\frac{7m}{10} = -7$

2 Solve the following equations.

a $2x + 9 = 15$ b $4x + 3 = 23$ c $3x + 5 = -4$

d $6b - 12 = 48$ e $-3b + 5 = 17$ f $-2b + 6 = 4$

g $-4k - 9 = 3$ h $-3k - 6 = -3$ i $6 - 5k = 16$

3 Solve the following equations.

a $\frac{x}{2} = 10$ b $\frac{4x}{7} = -4$ c $-\frac{2x}{3} = -6$

d $\frac{y}{2} - 5 = 3$ e $\frac{3y}{2} + 2 = 8$ f $\frac{3y}{5} - 1 = 8$

g $-\frac{4g}{7} - 1 = 11$ h $6 - \frac{4x}{5} = -2$ i $10 - \frac{7x}{9} = 3$

j $\frac{3x}{2} + 2 = 8$ k $5 - \frac{y}{2} = 1$ l $4 - \frac{2z}{3} = 0$

Challenge Yourself

1 Solve the following equations.

a $2a - \frac{3}{5} = 1$

b $\frac{a}{3} - 2 = \frac{1}{4}$

c $\frac{5k}{6} + 4 = -\frac{1}{3}$

d $\frac{7k}{9} - 1 = \frac{2}{3}$

e $-\frac{3p}{8} + \frac{1}{2} = \frac{5}{6}$

f $-\frac{2p}{3} - \frac{1}{9} = -\frac{3}{2}$

2 Solve for x in the equation $3x - p = q$

3 Solve for x in the equation $y = mx + c$

4 Solve for a in the equation $ab - c = d$

Variable on Both Sides: Moving, Collecting and Combining

Next, equations with the variable on both sides.

The goal's the same, solve for the variable.

Steps are the same too.

But now, since there's a variable on both sides,

you'll be moving things in both directions.

Left and right.

What you wanna do is get all the variable terms on one side.

And all the constants on the other.

It doesn't matter which side each goes to.

But to make it easier.

Check which side has the bigger coefficient.

That's the side to keep the variables on.

Why?

Because moving the smaller coefficient usually helps avoid negative numbers.

Once the variable terms are on one side, and the constants on the other.

Simplify each side by adding or subtracting terms.

This is called *combining like terms*.

After that, solve it like a basic equation.

So, here are the steps.

1) Move all the variable terms to one side
2) Move the constants to the other
3) Combine like terms
4) Solve for the variable

Worked Example 1

Solve the following linear equations.

a $\mathbf{3x + 7 = 2x - 1}$ b $\mathbf{5x - 3 = 2x + 9}$ c $\mathbf{3x + 2 = -x + 10}$

Explanation		Solution
a	Write the equation.	$3x + 7 = 2x - 1$
	Move variable terms to one side, constants to the other.	$3x - 2x = -1 - 7$
	Simplify.	$x = -8$
b	Write the equation.	$5x - 3 = 2x + 9$
	Move variable terms to one side, constants to the other.	$5x - 2x = 9 + 3$
	Simplify.	$3x = 12$
	Move the coefficient to the other side.	$x = 4$
c	Write the equation.	$3x + 2 = -x + 10$
	Move variable terms to one side, constants to the other.	$3x + x = 10 - 2$
	Simplify.	$4x = 8$
	Move the coefficient to the other side.	$x = 2$

Worked Example 2

Solve the following linear equations.

a $\boldsymbol{a - 5 = 2a - 6}$ b $\boldsymbol{17 - 3b = 3 - b}$ c $\boldsymbol{4m + 7 = -m + 8}$

Explanation	**Solution**
a Write the equation.	$a - 5 = 2a - 6$
Move variable terms to one side, constants to the other.	$-5 + 6 = 2a - a$
Simplify.	$1 = a \text{ or } a = 1$
b Write the equation.	$17 - 3b = 3 - b$
Move variable terms to one side, constants to the other.	$17 - 3 = -b + 3b$
Simplify.	$14 = 2b$
Move the coefficient to the other side.	$7 = b \text{ or } b = 7$
c Write the equation.	$4m + 7 = -m + 8$
Move variable terms to one side, constants to the other.	$4m + m = 8 - 7$
Simplify.	$5m = 1$
Move the coefficient to the other side.	$m = \frac{1}{5}$

Worked Example 3

Solve the following linear equation. $\frac{5x}{3}+\frac{2}{5}=\frac{x}{3}+\frac{9}{10}$

Explanation	Solution
Write the equation.	$\frac{5x}{3}+\frac{2}{5}=\frac{x}{3}+\frac{9}{10}$
Move variable terms to one side, constants to the other.	$\frac{5x}{3}-\frac{x}{3}=\frac{9}{10}-\frac{2}{5}$
Combine fractions and simplify.	$\frac{5x-x}{3}=\frac{9}{10}-\frac{4}{10}$
	$\frac{4x}{3}=\frac{5}{10}$
Move the coefficient to the other side.	$\frac{4x}{3}=\frac{1}{2}$
Simplify.	$x=\frac{1}{2}\times\frac{3}{4}$
	$x=\frac{3}{8}$

When moving things around both ways, take it one step at a time.
Don't try to do multiple moves in your head, even if it seems obvious.

Your Turn To Try

1 What is the value of x in the equation $3x + 5 = 2x + 12$?

A 7 B 6

C 5 D 4

2 Which value of y make this equation true $9y - 7 = 6y + 11$

A 2 B 6

C 3 D 4

3 What the value of z in the equation $7z - 3 = 5z + 13$

A 5 B 8

C 6 D 7

4 The solution to $4a + 8 = 2a + 18$ is?

A 6 B 4

C 5 D 7

5 What value of x makes this equation $8b + 3 = 4b + 15$ true?

A 4 B 1

C 2 D 3

Question	1	2	3	4	5
Answer	A	B	B	C	D

Time To Practice

1 Solve the following linear equations.

a $6 - x = 5 - 2x$

b $8 - 2x = 10 - 3x$

c $8x + 45 = 6x + 15$

d $7a + 20 = 10a - 10$

e $3 - g = 17 - 3g$

f $w + 8 = 14 - 5w$

g $15 - p = 27 - 3p$

h $14g = 2g + 60$

i $3h - 1 = 23 - h$

2 Simplify the linear following.

a $6w + 27 = 11w + 17$

b $5t + 4 = 10t - 11$

c $12r - 20 = 15r - 38$

d $20f - 4 = 7f + 35$

e $-4q - 30 = 10q + 110$

f $90 - 10v = 2 - 21v$

g $17 - 3n = 52 - 10n$

h $-5m - 9 = 27 - 11m$

i $2 - 6x = 10 - 4x$

j $-11t - 4 = 60 - 3t$

Challenge Yourself

1 Solve the following linear equations.

a $3x + 4 = 2x + \frac{5}{2}$

b $5x - \frac{3}{4} = 2x + \frac{1}{2}$

c $\frac{7q}{3} + 1 = \frac{5q}{3} + \frac{4}{5}$

d $\frac{8p}{5} - \frac{2}{3} = \frac{4p}{5} - \frac{1}{6}$

e $-\frac{s}{3} - \frac{1}{5} = -\frac{2s}{3} + \frac{3}{10}$

f $\frac{7t}{4} - \frac{1}{2} = \frac{5t}{4} + \frac{1}{3}$

Equations With Brackets: First Things First – Get Rid of 'Em

When solving linear equations with brackets.

First thing you do.

Get rid of the brackets.

Depending on how the equation's set up, there are two ways to do this.

We'll start with the first setup.

It's easier.

But it's less common.

It's when there's a bracket term on one side.

And just a number on the other.

But.

This is key.

The number in front of the brackets divides evenly into the number on the other side.

If it does.

Move the number in front of the brackets to the other side.

Using the opposite operation.

Once you've done that.

Remove the brackets.

From there, you're back to a basic equation.

Solve it like you did before.

If the number in front of the brackets doesn't go evenly into the number on the other side.

You can still move it.

You'll just end up with fractions.

In cases like this, it might be better to use the second way.

But we'll get into that after a couple of examples.

Worked Example 1

Solve the following linear equations.

a $3(x+4)=27$ b $5(2x-1)=15$ c $4(3x-2)=-44$

Explanation		Solution
a	Write the equation.	$3(x+4)=27$
	Move number outside of brackets to the other side.	$x+4=9$
	Move the constant to the other side and simplify.	$x=5$
b	Write the equation.	$5(2x-1)=15$
	Move number outside of brackets to the other side.	$2x-1=3$
	Move the constant to the other side.	$2x=4$
	Move the coefficient to the other side.	$x=2$
c	Write the equation.	$4(3x-2)=-44$
	Move number outside of brackets to the other side.	$3x-2=-11$
	Move the constant to the other side.	$3x=-9$
	Move the coefficient to the other side.	$x=-3$

When you move the constant or the coefficient, simplify the equation in one step if you can.

Worked Example 2

Solve the following linear equations.

a $-2(p+3)=16$ b $7(1-2q)=28$ c $-2(5r-1)=92$

	Explanation	Solution
a	Write the equation.	$-2(p+3)=16$
	Move number outside of brackets to the other side.	$p+3=-8$
	Move the constant to the other side and simplify.	$p=-11$
b	Write the equation.	$7(1-2q)=35$
	Move number outside of brackets to the other side.	$1-2q=5$
	Move the constant to the other side.	$-2q=4$
	Move the coefficient to the other side.	$q=-2$
c	Write the equation.	$-2(5r-1)=92$
	Move number outside of brackets to the other side.	$5r-1=-46$
	Move the constant to the other side.	$5r=-45$
	Move the coefficient to the other side.	$r=-9$

Equations With Brackets 2: First Things First – Distribute 'Em

Ok, that was a couple of examples of the first setup.

You probably won't see many of those.

What you'll mostly see is brackets on one side.

And on the other side?

It could be anything.

A number.

A variable.

Another bracket term.

Or a mix of all these.

But the idea's the same.

First step is to get rid of the brackets.

This time, you do that by *distributing.*

That just means multiply the number outside by each term inside the brackets.

Make sure to include the sign in front of each term when you multiply.

After multiplying, you'll either end up with a simple equation.

Or one that's got a variable on both sides.

From there, same as before.

Move things left and right, combine the variable and the constants.

Then solve.

Oh yeah.

If there's a minus sign in front of the brackets, with no number.

That's the same as having a -1.

So, multiply each term inside the brackets by -1.

And watch out for the positive and negative signs.

That's where most people mess up.

Worked Example 3

Solve the following linear equations.

a $\mathbf{2(4x-5)=7x}$ b $\mathbf{3(2x+8)=5x+15}$

Explanation		Solution
a	Write the equation.	$2(4x-5)=7x$
	Expand the brackets on the left side.	$8x-10=7x$
	Move variable terms to one side, constants to the other.	$8x-7x=10$
	Simplify.	$x=10$
b	Write the equation.	$3(2x+8)=5x+15$
	Expand the brackets on the left side.	$6x+24=5x+15$
	Move variable terms to one side, constants to the other.	$6x-5x=15-24$
	Simplify.	$x=-9$

Getting rid of brackets is also called *expanding* the brackets.

Worked Example 4

Solve the following linear equations.

a $2(a + 3) = 4(a - 1)$ b $-2(3b + 4) = -4(2b - 1)$

Explanation		Solution
a	Write the equation.	$2(a + 3) = 4(a - 1)$
	Expand the brackets on both sides.	$2a + 6 = 4a - 4$
	Move variable terms to one side, constants to the other.	$6 + 4 = 4a - 2a$
	Simplify.	$10 = 2a$
	Move the coefficient to the other side.	$a = 5$
b	Write the equation.	$-2(3b + 4) = -4(2b - 1)$
	Expand the brackets on both sides.	$-6b - 8 = -8b + 4$
	Move variable terms to one side, constants to the other.	$-6b + 8b = 4 + 8$
	Simplify.	$2b = 12$
	Move the coefficient to the other side.	$b = 6$

If there are brackets inside brackets, get rid of the inner ones first, simplify, then get rid on to the outer brackets.

Worked Example 5

Solve the following linear equations.

a $-2(4x-3)-7=3(2x+1)-5$ b $4(3x+2)+5=2(2x-1)+3$

Explanation	Solution
a Write the equation.	$-2(4x-3)-7=3(2x+1)-5$
Expand the brackets on both sides.	$-8x+6-7=6x+3-5$
Combine like terms on both sides.	$-8x-1=6x-2$
Move variable terms to one side, constants to the other.	$-1+2=6x+8x$
Simplify.	$1=14x$
Move the coefficient to the other side.	$x=\frac{1}{14}$
b Write the equation.	$4(3x+2)+5=2(2x-1)+3$
Expand the brackets on both sides.	$12x+8+5=4x-2+3$
Combine like terms on both sides.	$12x+13=4x+1$
Move variable terms to one side, constants to the other.	$12x-4x=1-13$
Simplify.	$8x=-12$
Move the coefficient to the other side.	$x=-\frac{12}{8}$
Simplify.	$x=-\frac{3}{2}$

Worked Example 6

Solve the following linear equations.

a $\mathbf{8(a-1)+2(a+2)+4=5}$ b $\mathbf{4(p-5(p-3))=0}$

Explanation	**Solution**
a Write the equation.	$8(a-1)+2(a+2)+4=5$
Expand the brackets on the left side.	$8a-8+2a+4+4=5$
Combine like terms on the left side.	$10a=5$
Move the coefficient to the other side.	$a=\frac{5}{10}$
Simplify.	$a=\frac{1}{2}$
b Write the equation.	$4(p-5(p-3))=0$
Move number outside of brackets to the other side.	$p-5(p-3)=0$
Expand the brackets on the left side.	$p-5p+15=0$
Combine like terms.	$-4p+15=0$
Move the constant to the other side.	$-4p=-15$
Move the coefficient to the other side.	$p=\frac{-15}{-4}$
Simplify.	$p=\frac{15}{4}$

Your Turn To Try

1 If $2(a+3)=10$, what does a equal?

A 3 B 2

C 5 D 4

2 What is the value of x in the equation $4(3x-2)=28$?

A 2 B 6

C 3 D 4

3 Solve for y and choose the correct answer if $2(y+5)=3y-1$.

A 12 B 14

C 11 D 13

4 Which of the following is the solution for z if $2(z-3)+3(z+2)=5$?

A 4 B 3

C 2 D 1

5 What is the solution for b given $5(3b+2)=2(7b+3)$?

A -4 B -1

C 4 D 1

Question	1	2	3	4	5
Answer	B	C	C	D	A

Time To Practice

1 Solve the following equations.

a $4(x+3)=16$ b $2(x-3)=12$ c $2(x-4)=18$

d $3(1-2p)=27$ e $4(2q-5)=-12$ f $2(5+4r)=-30$

g $3(m+5)=2m$ h $8(2-n)=-6n$ i $5(7-2s)=-15s$

2 Solve the following.

a $2(k-3)=k-5$ b $7(2-x)=8-x$

c $10-3c=2(4-c)$ d $4r+7=3(2r-5)$

e $2(3x-2)-3(x+1)=5$ f $3(2x+3)+2(x+4)=25$

g $2(2y-3)+3(4y-1)=23$ h $2(9-w)-3(w+16)=0$

i $7(j+1)-2(2j-3)=10$ j $7(2s-1)-4(s+3)=-49$

3 Solve the following.

a $5(u+3)=3(u+9)$ b $8(v-1)=4(v+3)$

c $9(2x-5)=3(4x+7)$ d $5(2y+9)=3(3y+13)$

e $5(l-3)=3(2l-1)$ f $2(4k+1)=5(3-k)$

g $4(2d+5)=3(d+15)$ h $3(4x-1)=7(2x-7)$

4 Solve the following.

a $4(2v-3)=3(3v+2)-5$ b $4(j+3)+2=2(3j+5)-8$

c $5(2w+3)-4=3(4w+5)-2$ d $2(x-4)+7x=3(x+2)+4x$

e $3(2m+1)-m=2(m+3)+2m$ f $8n-3(2n-4)=4(n+2)+2n$

Challenge Yourself

1 Solve the following linear equations.

a $3\left(\frac{3y}{5}-1\right)=15$

b $11\left(8-\frac{z}{4}\right)=99$

c $\frac{2}{3}(3k-6)-3(k+1)=11+k$

d $\frac{1}{2}(6z-8)+3(z+2)=4(2z-1)$

e $4-2(x-6)=\frac{2x}{3}$

f $\frac{7x}{5}-\frac{4x}{10}=2\left(x+\frac{9}{2}\right)$

g $2(h+5(h-1))=110$

h $-6(2t-3(t-4))=0$

i $2(2(p-3)+3(1-p))=0$

j $6-3(2-3(y+1))=18$

Equations With Fractions: Yep, We're Getting Rid of These Too

Just like with brackets, how you get rid of them.

With fractions, same.

You get rid of them too.

To get rid of a fraction.

You get rid of its denominator.

So, here's what to do.

Rewrite the equation as a single fraction.

Because when there's only one fraction,

there's only one denominator to get rid of.

To get rid of it.

Move it to the other side.

Using the opposite operation - multiplication.

To rewrite the equation as a single fraction.

Your fraction skills need to be solid.

You need to be able to combine fractions.

Fractions with fractions.

And fractions with numbers.

So, here's a quick refresher.

- A whole number can be turned into a fraction by writing it over 1 (e.g. $5 = \frac{5}{1}$).
- To add or subtract fractions, the denominators must be the same.
- To cross-multiply
    i. Multiply the top of the first fraction by the bottom of the second
    ii. Multiply the top of the second fraction by the bottom of the first

Equations with fractions show up in all sorts of ways.

Sometimes there's one fraction, sometimes two.

It could be on one side, or on both sides.

It might be mixed with numbers, or even inside brackets.

Whatever it is.

Doesn't matter.

What you want to do is rewrite it so there's a single fraction.

Either a single fraction on one side.

Or one on each side.

From there you move the denominators.

That'll leave you with a basic equation, or with the variable on both sides.

Back to what you already know.

Get rid of brackets, move things around, combine and solve.

Alright, before we get into examples.

This is the biggest mistake to avoid.

Sometimes the numerator will have more than one term.

Whenever you add, subtract, multiply or divide anything with more than one term.

Always, always, always put those terms inside brackets.

Especially if there are any negatives.

Why?

Because the operation applies to the *whole thing*.

Not just part of it.

Putting brackets around it makes that clear.

And it helps you avoid any sign mistakes too.

So yeah, watch out for those numerators.

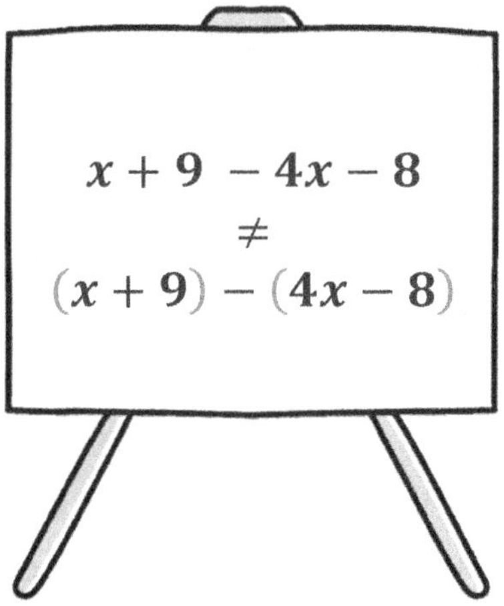

Here's a quick recap of the steps.

1) Rewrite as a single fraction on one or each side
2) Move the denominator - or denominators - to the other side
3) Get rid of any brackets
4) Move things around and combine like terms
5) Solve and simplify

Worked Example 1

Solve the following linear equations.

a $\frac{x}{3}+\frac{2}{3}=5$ b $\frac{11-5x}{4}=6$ c $\frac{5x-15}{3}=2x$

Explanation	Solution
a Write the equation.	$\frac{x}{3}+\frac{2}{3}=5$
The denominators are the same so add the numerators. (making left side one fraction)	$\frac{x+2}{3}=5$
Move the denominator to the other side.	$x+2=15$
Move the constant to the other side.	$x=13$
b Write the equation.	$\frac{11-5x}{4}=-6$
Move the denominator to the other side.	$11-5x=-24$
Move the constant to the other side.	$-5x=-35$
Move the coefficient to the other side and simplify.	$x=7$
c Write the equation.	$\frac{5x-15}{3}=2x$
Move the denominator to the other side.	$5x-15=6x$
Move variable terms to one side and constants to the other.	$-15=6x-5x$ $-15=x$
Simplify.	$x=-15$

Worked Example 2

Solve the following linear equations.

a $\quad 4+\dfrac{x-5}{2}=-3$

b $\quad 1-\dfrac{2-x}{3}=2$

Explanation	**Solution**
a Write the equation.	$4+\dfrac{x-5}{2}=-3$
Move the constant to the right side. (making left side one fraction)	$\dfrac{x-5}{2}=-7$
Move the denominator to the other side.	$x-5=-14$
Move the constant to the other side.	$x=-9$
b Write the equation.	$1-\dfrac{2-x}{3}=2$
Move the constant to the right side. (making left side one fraction)	$-\dfrac{2-x}{3}=1$
Move the negative to the other side.	$\dfrac{2-x}{3}=-1$
Move the denominator to the other side.	$2-x=-3$
Move the constant to the other side.	$-x=-5$
Simplify.	$x=5$

If there's a negative in front of a fraction, it's same as being multiplied by −1.
Move it by dividing the other side by −1.

Worked Example 3

Solve the following linear equations.

a $\frac{5x}{7} - \frac{x}{14} = \frac{9}{7}$ b $\frac{x}{4} - \frac{x}{5} = \frac{3}{4}$

Explanation	Solution
a Write the equation.	$\frac{5x}{7} - \frac{x}{14} = \frac{9}{7}$
Multiply top and bottom of first fraction by a number to match second denominator.	$\frac{10x}{14} - \frac{x}{14} = \frac{9}{7}$
Subtract fractions on the left side. (making left side one fraction)	$\frac{10x - x}{14} = \frac{9}{7}$
	$\frac{9x}{14} = \frac{9}{7}$
Move the denominator to the other side and simplify.	$9x = \frac{9}{7} \times 14$
Move the coefficient to the other side.	$9x = 18$
Simplify.	$x = 2$
b Write the equation.	$\frac{x}{4} - \frac{x}{5} = \frac{3}{4}$
Subtract fractions on the left side by cross multiplying. (making left side one fraction)	$\frac{5x - 4x}{20} = \frac{3}{4}$
Simplify numerator.	$\frac{x}{20} = \frac{3}{4}$
Move the denominator to the other side.	$x = \frac{3}{4} \times 20$
Simplify.	$x = 15$

Worked Example 4

Solve the following linear equation. $\frac{a+1}{3}+\frac{a+5}{5}=4$

Explanation	**Solution**
Write the equation.	$\frac{a+1}{3}+\frac{a+5}{5}=4$
Add fractions on the left side by cross multiplying. (making left side one fraction)	$\frac{5(a+1)+3(a+5)}{15}=4$
Expand brackets in the numerator.	$\frac{5a+5+3a+15}{15}=4$
Combine like terms in the numerator.	$\frac{8a+20}{15}=4$
Move the denominator to the other side.	$8a+20=60$
Move the constant to the other side.	$8a=40$
Move the coefficient to the other side.	$a=5$

When combining fractions, simplify as you go.
This makes the equation easier to solve.

Worked Example 5

Solve the following linear equation. $\frac{b+1}{2} - \frac{b-7}{6} = 3$

Explanation	Solution
Write the equation.	$\frac{b+1}{2} - \frac{b-7}{6} = 3$
Multiply top and bottom of first fraction by a number to match second denominator.	$\frac{3(b+1)}{3 \times 2} - \frac{b-7}{6} = 3$
The denominators are the same so add the numerators. (making left side one fraction)	$\frac{3(b+1)-(b-7)}{6} = 3$
Expand brackets in the numerator.	$\frac{3b+3-b+7}{6} = 3$
Combine like terms in the numerator.	$\frac{2b+10}{6} = 3$
Move the denominator to the other side.	$2b + 10 = 18$
Move the constant to the other side.	$2b = 8$
Move the coefficient to the other side.	$b = 4$

You can also cross multiply when there's one fraction on each side of the equals sign. As long as nothing's added or subtracted to them.

Worked Example 6

Solve the following linear equations.

a $\dfrac{x+1}{3} = \dfrac{x-1}{5}$

b $\dfrac{2x-3}{2} = \dfrac{-x-1}{4}$

Explanation	Solution
a Write the equation.	$\dfrac{x+1}{3} = \dfrac{x-1}{5}$
Cross multiply.	$5(x+1) = 3(x-1)$
Expand the brackets on both sides.	$5x+5 = 3x-3$
Move variable terms to one side, constants to the other.	$5x-3x = -3-5$
Move the coefficient to the other side.	$2x = -8$
Simplify.	$x = -4$
b Write the equation.	$\dfrac{2x-3}{2} = \dfrac{-x-1}{4}$
Cross multiply.	$4(2x-3) = 2(-x-1)$
Expand the brackets on both sides.	$8x-12 = -2x-2$
Move variable terms to one side, constants to the other.	$10x = -2+12$
Move the coefficient to the other side.	$10x = 10$
Simplify.	$x = 1$

Worked Example 7

Solve the following linear equation. $\frac{2y-1}{3}+5=\frac{y+2}{4}$

Explanation	Solution
Write the equation.	$\frac{2y-1}{3}+5=\frac{y+2}{4}$
Rewrite the constant on the left side as a fraction.	$\frac{2y-1}{3}+\frac{5}{1}=\frac{y+2}{4}$
Add fractions on the left side by cross multiplying. (making left side one fraction)	$\frac{(2y-1)+15}{3}=\frac{y+2}{4}$
Expand brackets in the numerator on the left side.	$\frac{2y-1+15}{3}=\frac{y+2}{4}$
Combine like terms in the numerator.	$\frac{2y+14}{3}=\frac{y+2}{4}$
Cross multiply.	$4(2y+14)=3(y+2)$
Expand the brackets on both sides.	$8y+56=3y+6$
Move variable terms to one side, constants to the other.	$8y-3y=6-56$
Move the coefficient to the other side.	$5y=-50$
Simplify.	$y=-10$

When working with two or more terms not writing them inside brackets. ⚠

Worked Example 8

Solve the following linear equation. $\frac{y+2}{3}-\frac{y-1}{2}=\frac{y-7}{4}$

Explanation	Solution
Write the equation.	$\frac{y+2}{3}-\frac{y-1}{2}=\frac{y-7}{4}$
Subtract fractions on the left side by cross multiplying. (making left side one fraction)	$\frac{2(y+2)-3(y-1)}{6}=\frac{y-7}{4}$
Expand brackets in the numerator on the left side.	$\frac{2y+4-3y+3}{6}=\frac{y-7}{4}$
Combine like terms in the numerator.	$\frac{-y+7}{6}=\frac{y-7}{4}$
Cross multiply.	$4(-y+7)=6(y-7)$
Expand the brackets on both sides.	$-4y+28=6y-42$
Move variable terms to one side, constants to the other.	$28+42=6y+4y$
Move the coefficient to the other side.	$70=10y$
Simplify.	$y=\frac{70}{10}$ $y=7$

Worked Example 9

Solve the following linear equation. $\frac{3r-5}{2} - \frac{4r-7}{3} = \frac{4-2r}{4}$

Explanation	Solution
Write the equation.	$\frac{3r-5}{2} - \frac{4r-7}{3} = \frac{4-2r}{4}$
Subtract fractions on the left side by cross multiplying. (making left side one fraction)	$\frac{3(3r-5) - 2(4r-7)}{6} = \frac{4-2r}{4}$
Expand brackets in the numerator on the left side.	$\frac{9r - 15 - 8r + 14}{6} = \frac{4-2r}{4}$
Combine like terms in the numerator.	$\frac{r-1}{6} = \frac{4-2r}{4}$
Cross multiply.	$4(r-1) = 6(4-2r)$
Expand the brackets on both sides.	$4r - 4 = 24 - 12r$
Move variable terms to one side, constants to the other.	$4r + 12r = 24 + 4$
Move the coefficient to the other side.	$16r = 28$
Simplify.	$r = \frac{28}{16}$
	$r = \frac{7}{4}$

Your Turn To Try

1 **What is the value of x in the equation $\frac{x}{2}+\frac{3x}{2}=16$?**

A 2   B 4

C 8   D 6

2 **Find the value of x in the equation $\frac{4x}{3}+\frac{x}{3}=10$?**

A 2   B 5

C 3   D 6

3 **Which value of y satisfies $\frac{2y+1}{3}=y-2$?**

A 3   B 7

C 5   D 9

4 **What is the solution to $\frac{2a+1}{3}=\frac{a+4}{2}$?**

A -8 B 8

C -10 D 10

5 **Find the value of b in the equation $\frac{3b+4}{4}=\frac{3b}{2}-2$.**

A 0   B 3

C 4   D 1

Question	1	2	3	4	5
Answer	C	D	B	D	C

Time To Practice

1 Solve the following linear equations.

a $\frac{x+1}{2} = 3$ b $\frac{y-3}{3} = 4$ c $\frac{z-9}{3} = -1$

d $\frac{3x+5}{2} = 7$ e $\frac{5y-12}{3} = 11$ f $\frac{3z-4}{7} = -4$

g $\frac{7x-12}{3} = -25$ h $\frac{29-2y}{3} = 5$ i $\frac{100-5z}{3} = 30$

j $\frac{8x+4}{5} = 2x$ k $\frac{12-4y}{6} = -y$ l $\frac{15-z}{3} = -2z$

2 Solve the following linear equations.

a $\frac{3x-1}{4} + 6 = 5$ b $\frac{12-3y}{3} - 3 = 5$ c $\frac{35-2z}{7} - 3 = 2$

d $\frac{7x+4}{3} - 8 = -9$ e $\frac{3y+1}{4} - 3 = -5$ f $\frac{4z-6}{3} - 7 = 3$

3 Solve the following linear equations.

a $\frac{x}{3} - \frac{x}{2} = 1$ b $\frac{y}{4} - \frac{y}{5} = 1$ c $\frac{5z}{4} - \frac{z}{2} = -\frac{3}{4}$

d $\frac{x}{4} - \frac{x}{7} = \frac{9}{14}$ e $y + \frac{y}{7} = 8$ f $-z - \frac{z}{4} = \frac{5}{4}$

g $2x + \frac{3x}{5} = -13$ h $6y - \frac{3y}{2} = 27$ i $-4z + \frac{5z}{2} = -15$

4 Solve the following linear equations.

a $\frac{2x+1}{3} = x - 2$ b $\frac{5y-3}{2} = 2y + 4$ c $\frac{z+17}{5} = z + 1$

d $4x - 9 = \frac{2x+3}{3}$ e $y + 3 = \frac{9-y}{2}$ f $-3z - 2 = \frac{8z-33}{5}$

Time To Practice

5 Solve the following linear equations.

a $\frac{3x+7}{5}=\frac{2x}{3}$

b $\frac{3y-10}{7}=\frac{2y}{5}$

c $\frac{z-8}{3}=\frac{z}{6}$

d $\frac{x-5}{3}=\frac{x-7}{4}$

e $\frac{y+2}{6}=\frac{y-5}{3}$

f $\frac{z+3}{4}=\frac{27-z}{6}$

g $\frac{9x+3}{2}=\frac{13x+7}{3}$

h $\frac{4y+9}{15}=\frac{y+2}{5}$

i $\frac{2z+7}{5}=\frac{3-z}{4}$

j $\frac{x+11}{2}=\frac{x+14}{3}$

k $\frac{3-z}{2}=\frac{6(9-z)}{4}$

l $\frac{3y-4}{2}=\frac{102-5y}{4}$

6 Solve the following linear equations.

a $\frac{x+3}{2}-3=\frac{x+4}{4}$

b $\frac{x+3}{10}=2-\frac{5-x}{2}$

c $\frac{x-1}{4}=\frac{x+3}{2}-4$

d $\frac{4x+9}{15}=1+\frac{x-3}{5}$

e $\frac{6x+7}{5}=\frac{5x+1}{4}+1$

f $\frac{2x+10}{3}=\frac{6-5x}{4}-2$

g $\frac{3x-4}{5}=\frac{5x+9}{7}-3$

h $6-\frac{7-2x}{3}=\frac{1-2x}{9}$

7 Solve the following.

a $\frac{5x-10}{3}-\frac{3x}{2}=\frac{2x}{3}$

b $\frac{-3x+12}{6}+\frac{2x-6}{3}=-\frac{x}{2}$

c $\frac{2x-4}{3}+\frac{x+5}{6}=\frac{2-x}{4}$

d $\frac{-x-6}{5}+\frac{4x+7}{10}=\frac{3x+1}{5}$

e $\frac{5x+8}{5}-\frac{3x+6}{10}=\frac{x+1}{2}$

f $\frac{3x-8}{3}+\frac{3-x}{6}=\frac{2x-5}{4}$

Challenge Yourself

1 Solve the following equations.

a $\frac{7x}{5} - \frac{4x}{10} = 2\left(x + \frac{9}{2}\right)$

b $\frac{3}{4} - 2\left(a + \frac{1}{5}\right) = \frac{a}{3}$

c $\frac{41}{9} = \frac{5}{2}\left(y + \frac{2}{3}\right) - \frac{y}{3}$

d $-k - \frac{5}{4}\left(1 - \frac{8k}{3}\right) = -\frac{19}{4}$

e $\frac{1}{2}\left(1 - \frac{2n}{3}\right) = \frac{n}{5} + \frac{1}{2}$

f $\frac{7}{6} - \frac{4p}{3} = 2\left(p + \frac{3}{2}\right) - \frac{3p}{2}$

4. Answers

Basic Equations

Time to Practice

1 a 2 b 11 c 5
d -7 e 7 f -11
g 12 h 20 i -10

2 a 3 b 5 c -3
d 10 e -4 f 1
g -3 h -1 i -2

3 a 20 b -7 c 9
d 16 e 4 f 15
g -21 h 10 i 9
j 4 k 8 l 6

Challenge Yourself

1 a $\frac{4}{5}$ b $\frac{27}{4}$ c $-\frac{26}{5}$
d $\frac{15}{7}$ e $-\frac{8}{9}$ f $\frac{25}{12}$

2 $x = \frac{q+p}{3}$

3 $x = \frac{y-c}{m}$

4 $a = \frac{d+c}{b}$

Variables on Both Sides

Time to Practice

1 a -1 b 2 c -15
d 10 e 7 f 1
g 6 h 5 i 6

2 a 2 b 3 c 6
d 3 e -10 f -8
g 5 h 6 i -4
j -8

Challenge Yourself

1 a $-\frac{3}{2}$ b $\frac{5}{12}$ c $-\frac{3}{10}$
d $\frac{5}{8}$ e $\frac{3}{2}$ f $\frac{5}{3}$

Equations With Brackets

Time to Practice

1	a	1	b	9	c	13
	d	−4	e	1	f	−5
	g	−15	h	8	i	−7
2	a	1	b	1	c	2
	d	11	e	4	f	1
	g	2	h	−6	i	−1
	j	−3				
3	a	6	b	5	c	11
	d	−6	e	−12	f	1
	g	5	h	23		
4	a	−13	b	6	c	−1
	d	7	e	3	f	1

Challenge Yourself

1	a	10	b	−4	c	−9
	d	3	e	6	f	−9
	g	10	h	12	i	−3
	j	1				

Equations With Fractions

Time to Practice

1	a 5	b 15	c 6	d 3	e 9
	f −8	g −9	h 7	i 2	j 2
	k −6	l −3			
2	a −1	b −4	c 0	d −1	e −3
	f 9				
3	a −6	b 20	c −1	d 6	e 7
	f −1	g −5	h 6	i 10	
4	a 7	b 11	c 3	d 3	e 1
	f 1				
5	a 21	b 50	c 16	d −1	e 12
	f 9	g 5	h −3	i −1	j −5
	k 12	l 10			
6	a 10	b 2	c 9	d −3	e 3
	f −2	g 8	h −4		
7	a $-\frac{20}{3}$	b 0	c $\frac{12}{13}$	d $-\frac{7}{4}$	e $-\frac{5}{2}$
	f $\frac{11}{4}$				

Challenge Yourself

1	a −9	b $\frac{3}{20}$	c $\frac{4}{3}$	d $-\frac{3}{2}$	e 0
	f −1				

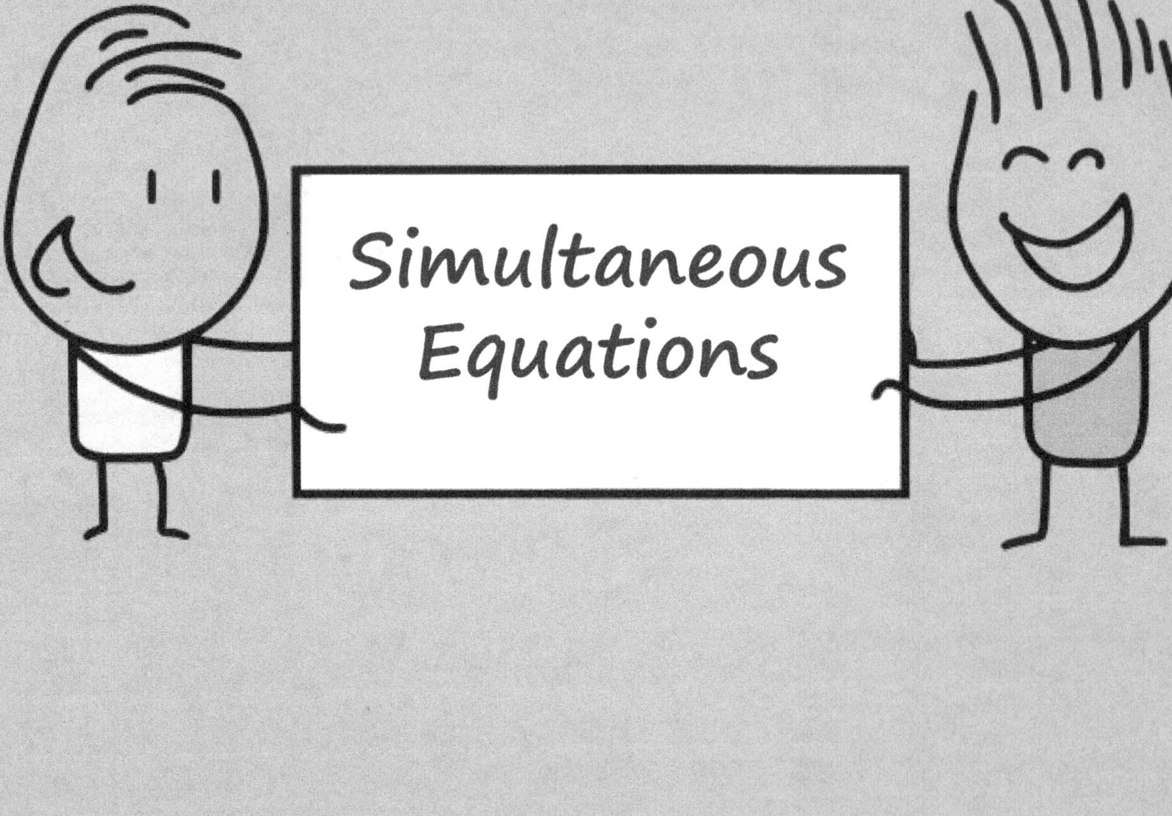
Simultaneous
Equations

Simultaneous Equations: Double the Equations, Double the Fun (Sort Of)

You're good with linear equations.

Now it's time to look at simultaneous equations.

A linear equation has one equation with one variable.

Simultaneous equations have two equations with two variables.

Since there are two variables, the answer will be two numbers.

One for each variable.

And if you plug those numbers into both equations, they should both work.

Meaning the left side of each equation will equal the right side.

If the two numbers only work for one equation, the answer's wrong.

It has to work for *both* equations.

To solve for two variables, you start by solving for one.

Once you solve the first, the second one is solved like the linear equations you did earlier.

There are two ways to solve for the first variable.

The Substitution Method.

The Elimination Method.

How do you know which to use?

Sometimes, you'll be told which one to use.

If not, it depends on the coefficients.

One method is usually easier than the other.

We'll go through both.

Once you see how each method works,

you'll know which one to use as soon as you look at the question.

The Substitution Method: Spot a 1 or −1 and You're Good to Go

'Substitution' means replacing one thing with something else.

That's the basic idea of this method.

You replace a variable with an expression.

This substitution is what lets you solve for the first variable.

Once you've got the first, you can then solve for the second.

Here's what you do.

Look at the coefficients of each equation.

What you're looking for is a variable with a coefficient of 1 or −1.

Preferably 1.

But −1 is fine too.

Then rearrange the equation with that variable to make it by itself on one side.

When you do this, you'll get a variable equal to an expression.

That expression will be in terms of the other variable.

From here, in the other equation.

The one you didn't rearrange.

Replace this variable with the expression it equals.

This gives you an equation with just one variable.

Solve it.

That's the first variable done.

Now plug that value back into the equation you rearranged.

And solve for the second variable.

That's pretty much it.

Alright.

Just a couple of quick points.

Always label the equations.

The originals as (1) and (2).

The new one as (3).

This makes it easier to refer back to the equations.

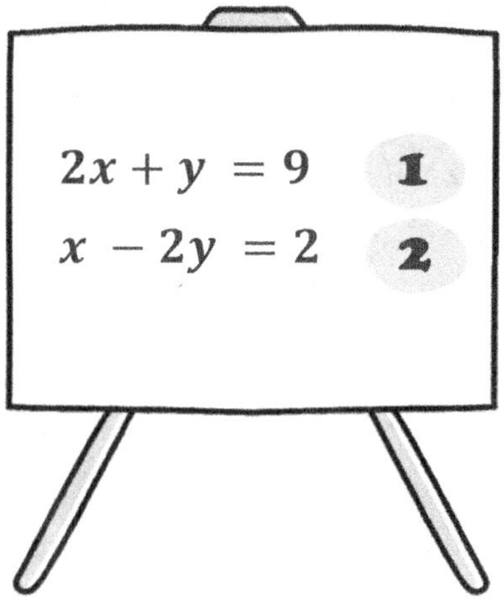

Also, as you're working it out.

Write a short note about what you're doing.

Just like the ones you'll see in the examples.

That way, anyone reading your work, like your teacher, can follow along.

Here are the 6 steps for the Substitution Method.

1) Pick a variable with a coefficient of 1 or -1
2) Rearrange the equation to get that variable by itself
3) Substitute the expression from Step 2 into the other equation
4) Solve the new equation
5) Plug the solution back into one of the original equations to find the other variable
6) Write the answers

One last thing.

If the question specifically asks to use the Substitution Method,

but none of the coefficients are 1 or -1.

No stress.

Follow the same steps.

It'll still work.

You'll just be working with fractions.

Worked Example 1

Use the substitution method to solve the following simultaneous equations.

$$\mathbf{2x + y = 5}$$
$$\mathbf{x - y = 1}$$

Explanation	**Solution**
Write and label the equations.	$2x + y = 5 \quad (1)$ $x - y = 1 \quad (2)$
Choose to make x the subject in the second equation. (move y to the other side)	$x - y = 1$ $x = 1 + y \quad (3)$
Substitute (or replace) x in the first equation with this expression.	**Sub (3) into (1)** $2x + y = 5$ $2(1 + y) + y = 5$
Solve this new equation for the variable y. (expand brackets on the left, collect like terms and solve)	$2 + 2y + y = 5$ $3y + 2 = 5$ $3y = 3$ $y = 1$
Plug this value into either original equations and solve for x. (e.g. second equation)	**Sub $y = 1$ into (2)** $x - y = 1$ $x - 1 = 1$ $x = 2$
Write answer.	$\therefore x = 2 \text{ and } y = 1$

If one equation already has a variable as the subject, substitute that expression directly into the other equation.

Worked Example 2

Use the substitution method to solve the following simultaneous equations.

$$\boldsymbol{y = x + 3}$$
$$\boldsymbol{2x - 3y = -8}$$

Explanation	Solution
Write and label the equations.	$y = x + 3 \quad (1)$ $2x - 3y = -8 \quad (2)$
Substitute (or replace) y in the second equation with given expression.	Sub (1) into (2) $2x - 3(x + 3) = -8$
Solve this new equation for the variable x. (expand the brackets of the left, collect like terms and solve)	$2x - 3x - 9 = -8$ $-x - 9 = -8$ $-x = 1$ $x = -1$
Plug this value into either original equations and solve for y. (e.g. first equation).	Sub $x = -1$ into (1) $y = -1 + 3$ $y = 2$
Write answer.	$\therefore x = -1 \text{ and } y = 2$

Always use brackets around the substituted expression to avoid mistakes with negative signs or multiplication.

Worked Example 3

Use the substitution method to solve the following simultaneous equations.

$$\boldsymbol{a + 2b = 8}$$
$$\boldsymbol{2a - b = 1}$$

Explanation	Solution
Write and label the equations.	$a + 2b = 8 \quad (1)$ $2a - b = 1 \quad (2)$
Choose to make a the subject in the first equation. (move $2b$ to the other side)	$a + 2b = 8$ $a = 8 - 2b \quad (3)$
	Sub (3) into (2)
Substitute (or replace) a in the second equation with this expression.	$2a - b = 1$ $2(8 - 2b) - b = 1$
Solve this new equation for the variable b. (expand brackets on the left, collect like terms and solve)	$16 - 4b - b = 1$ $16 - 5b = 1$ $5b = 15$ $b = 3$
	Sub $b = 3$ into (1)
Plug this value into either equation and solve for a. (e.g. first equation)	$a + 2b = 8$ $a + 2(3) = 8$ $a + 6 = 8$ $a = 2$
Write answer.	$\therefore a = 2 \text{ and } b = 3$

If both variables have a coefficient of 1 or -1, go with the one that has positive 1. It faster, easier to solve and avoids using negatives.

Worked Example 4

Use the substitution method to solve the following simultaneous equations.

$$5m - 2n = 18$$
$$-2m - n = -9$$

Explanation	Solution
Write and label the equations.	$5m - 2n = 18 \quad (1)$ $-2m - n = -9 \quad (2)$
Choose to make n the subject in the second equation. (rearrange)	$-2m - n = -9$ $n = 9 - 2m \quad (3)$
	Sub (3) into (1)
Substitute (or replace) n in the first equation with this expression.	$5m - 2n = 18$ $5m - 2(9 - 2m) = 18$
Solve this new equation for the variable m. (expand brackets on the left, collect like terms and solve)	$5m - 18 + 4m = 18$ $9m - 18 = 18$ $9m = 36$ $m = 4$
	Sub $m = 4$ into (2)
Plug this value into either original equations and solve for n. (e.g. second equation)	$-2m - n = -9$ $-2(4) - n = -9$ $-8 - n = -9$ $-8 + 9 = n$ $n = 1$
Write answer.	$\therefore m = 4 \; and \; n = 1$

Worked Example 5

Solve the simultaneous equations $2x + 3y = 5$ and $x - y = \frac{1}{2}$

Explanation	Solution
Write and label the equations.	$2x + 3y = 5 \quad (1)$ $x - y = \frac{1}{2} \quad (2)$
Choose to make x the subject in the second equation. (move y to the other side)	$x - y = \frac{1}{2}$ $x = \frac{1}{2} + y \quad (3)$
	Sub (3) into (1)
Substitute (or replace) x in the first equation with this expression.	$2x + 3y = 5$ $2\left(\frac{1}{2} + y\right) + 3y = 5$
Solve this new equation for the variable y. (expand brackets on the left, collect like terms and solve)	$1 + 2y + 3y = 5$ $1 + 5y = 5$ $5y = 4$ $y = \frac{4}{5}$
	Sub $y = \frac{4}{5}$ into (2)
Plug this value into either original equations and solve for x. (e.g. second equation)	$x - y = \frac{1}{2}$ $x - \frac{4}{5} = \frac{1}{2}$ $x = \frac{1}{2} + \frac{4}{5} = \frac{13}{10}$
Write answer.	$\therefore x = \frac{13}{10}\ and\ y = \frac{4}{5}$

Your Turn To Try

1 **For the following equations. What is the value of x?**

$$\mathbf{y = x + 5}$$
$$\mathbf{3x + y = 13}$$

A $x = 3$ B $x = 2$ C $x = 1$ D $x = 4$

2 **Solving the following equations. What is the value of y?**

$$\mathbf{y = x + 3}$$
$$\mathbf{6x + y = 17}$$

A $y = 2$ B $y = 4$ C $y = 5$ D $x = 6$

3 **Find the value of x after using the substitution method.**

$$\mathbf{x = y - 1}$$
$$\mathbf{2x + 3y = 8}$$

A $x = 3$ B $x = 2$ C $x = 1$ D $x = 4$

4 **Given the following equations. What is the value of a?**

$$\mathbf{4a + 5b = 6}$$
$$\mathbf{a - 2b = -5}$$

A $a = 2$ B $a = -1$ C $a = 1$ D $a = -2$

5 **After solving the following simultaneous equations. What is the solution for n?**

$$\mathbf{3m + n = 5}$$
$$\mathbf{5m - 4n = -3}$$

A $n = -1$ B $n = -2$ C $n = 1$ D $n = 2$

Question	1	2	3	4	5
Answer	B	C	C	B	D

Time To Practice

1 Solve the following pairs of simultaneous equations, using the substitution method.

a $y = x$ and $4x + 3y = 7$

b $y = x$ and $7x + 3y = 20$

c $y = x - 1$ and $3x + 2y = 8$

d $y = x + 1$ and $3x - 2y = -5$

e $x = 2y + 1$ and $11y - 5x = -2$

f $x = 3y - 3$ and $7y - 2x = 11$

g $x = 2y + 12$ and $3x + 2y = -4$

h $x = 4y + 2$ and $2y - 3x = 14$

2 Solve the following pairs of simultaneous equations, using the substitution method.

a $3x + 4y = 27$
$2x - y = -4$

b $3x - y = 23$
$2x - 2y = 10$

c $4a + 3b = 23$
$2a - b = -1$

d $2a + b = 16$
$3a - 4b = 24$

e $5p - q = -22$
$7p + 2q = -7$

f $5p + 2q = 5$
$3p - q = 25$

g $3m - 2n = 34$
$m + 4n = 16$

h $6m - 3n = 21$
$2m + n = -15$

i $5r + 3s = 44$
$2r - s = 0$

j $4r + 3s = -59$
$r - 2s = 10$

Challenge Yourself

1 Solve the following simultaneous equations using the substitution method.

a $m - 2n - 5 = 0$
$3m - n - 25 = 0$

b $y + x - 10 = 0$
$7y + 2x - 5 = 0$

c $r - 2s = 4$
$r + \frac{s}{3} = 4$

d $2c + \frac{d}{4} = 7$
$c - \frac{d}{2} = 1$

e $3p + q = 5$
$p - \frac{q}{3} = 1$

f $2a - b = 3$
$a + \frac{b}{2} = 2$

The Elimination Method: Line 'Em Up, Knock One Out

'Elimination' means getting rid of something.

And that's pretty much what we're doing in this method.

We're getting rid of one of the variables.

Once one's gone, we can solve for the other.

This method even works when the coefficients are 1 or -1.

But it's when they're not that this method is the better choice to use.

Here's what to do.

First thing.

Very important.

The equations need to be written right on top of each other.

Vertically aligned.

The variables, numbers, and equal signs all lined up neatly on top of each other.

Sometimes they already come lined up.

If they are, sweet.

If not, rewrite them and line them up.

It makes the elimination step a lot easier.

Once you've got the equations lined up.

Label them as (1) and (2).

From here, there are three different scenarios.

We'll go through them one by one.

First scenario.

This is when one variable has the same coefficient in *both* equations.

The sign doesn't matter.

It can be positive or negative.

We just want the same number in front of one variable.

That's the variable we'll eliminate first.

How do you eliminate it?

You add or subtract the equations.

And you do this vertically.

Add or subtract the variables and the numbers.

That's why we line up the equations right on top of each other.

This gets rid of the variable with the same coefficient.

Leaving a new equation with just one variable.

Solve that.

Then plug that answer into one of the original equations to solve for the second variable.

Here's the part to watch out for.

How do you know whether to add or subtract the equations?

Pick the wrong one, and no variable gets eliminated.

Easy.

Here's how.

- If the signs are the same, subtract.
- If the signs are different, add.

Alright, let's wrap this up.

Solving by elimination.

Scenario 1.

Here are the steps.

1) Line up and label the equations (1) and (2)
2) Pick a variable with the same coefficient
3) Add or subtract the equations to eliminate that variable
4) Solve the new equation for the remaining variable
5) Plug this answer back into either original equation to solve for the other variable

Worked Example 1

Solve the following simultaneous equations using the elimination method.

$$3x + 5y = 11$$
$$2x + 5y = 9$$

Explanation	Solution
Write and label the equations.	$3x + 5y = 11 \quad (1)$ $2x + 5y = 9 \quad (2)$
	$(1) - (2)$
Choose the variable y to eliminate. Subtract the second equation from the first.	$3x - 2x + 5y - 5y = 11 - 9$ $x = 2$
	Sub $x = 2$ into (2)
Plug this value into either original equations and solve for y. (e.g. second equation)	$2x + 5y = 9$ $2(2) + 5y = 9$ $4 + 5y = 9$
	$5y = 9 - 4$ $5y = 5$ $y = 1$
Write answer.	$\therefore x = 2 \text{ and } y = 1$

When adding or subtracting the equations, add or subtract down each column. Variables first, then numbers.

Worked Example 2

Solve the following simultaneous equations using the elimination method.

$$\mathbf{6x + 4y = -10}$$
$$\mathbf{-6x + 5y = 1}$$

Explanation	Solution
Write and label the equations.	$6x + 4y = -10 \quad (1)$ $-6x + 5y = 1 \quad (2)$
	$(2) + (1)$
Choose the variable x to eliminate. Add the first equation to the second.	$-6x + 6x + 4y + 5y = -10 + 1$ $9y = -9$
Solve this new equation for y.	$y = -9 \div 9$ $y = -1$
	Sub $y = -1$ into (1)
Plug this value into either original equations and solve for x. (e.g. first equation)	$6x + 4y = -10$ $6x + 4(-1) = -10$ $6x - 4 = -10$
	$6x = -10 + 4$ $6x = -6$ $x = -1$
Write answer.	$\therefore x = -1 \ and \ y = -1$

When adding or subtracting the equations, don't write the variable being eliminating. It'll be zero anyway.

Worked Example 3

Solve the following simultaneous equations using the elimination method.

$$\mathbf{3x - 2y = 5}$$
$$\mathbf{-4x - 2y = -2}$$

Explanation	Solution
Write and label the equations.	$3x - 2y = 5 \quad (1)$ $-4x - 2y = -2 \quad (2)$
	$\mathbf{(1) - (2)}$
Choose the variable y to eliminate. Subtract the second equation from the first.	$3x - (-4x) = 5 - (-2)$ $3x + 4x = 5 + 2$ $7x = 7$
Solve this new equation for x.	$x = 7 \div 7$ $x = 1$
	Sub $x = 1$ into (1)
Plug this value into either original equations and solve for y. (e.g. first equation)	$3x - 2y = 5$ $3(1) - 2y = 5$ $3 - 2y = 5$ $-2y = 2$ $y = -1$
Write answer.	$\therefore x = 1 \text{ and } y = -1$

When solving for the second variable, plug back into the simpler equation. Usually, the one with fewer negatives or smaller numbers.

Worked Example 4

Solve the following simultaneous equations using the elimination method.

$$5x = 17 - 2y$$
$$2y = 3x - 7$$

Explanation	Solution
Line up and label the equations.	$5x + 2y = 17 \quad (1)$ $3x - 2y = 7 \quad (2)$
	$(2) + (1)$
Choose the variable y to eliminate. Add the first equation to the second.	$3x + 5x = 7 + 17$ $8x = 24$
Solve this new equation for x.	$x = 24 \div 8$ $x = 3$
	Sub $x = 3$ into (2)
Plug this value into either original equations and solve for y. (e.g. second equation)	$3x - 2y = 7$ $3(3) - 2y = 7$ $9 - 2y = 7$ $-2y = -2$ $y = 1$
Write answer.	$\therefore x = 3 \text{ and } y = 1$

It doesn't matter which equation is labelled (1) or (2). What matters is the sign of the coefficients. Add when different. Subtract when same.

The Elimination Method 2: Matching the Coefficients

The Elimination Method only works when one variable has the same coefficient.

In *both* equations.

If it doesn't, we change the equations so it does.

Remember, for the coefficient, the $+$ or $-$ sign doesn't matter.

As long as the number is the same.

Okay, here's the second scenario.

The coefficients are not the same.

But.

For one of the variables,

the smaller coefficient goes evenly into the bigger one.

That's the variable to eliminate.

To do that, we first need to make the coefficients the same.

How?

Figure out what to multiply the smaller coefficient by to get the bigger one.

(Hint. Divide the bigger coefficient by the smaller one.)

Then multiply the entire equation with the smaller coefficient by that number.

Make sure you multiply everything.

Each variable and the number on the right side of the equal sign.

This gives you a new equation.

Where the coefficients of that variable now match one of the original equations.

Label this new equation as (3).

From here you're back to scenario one.

Line up the two equations with the same coefficient.

Add or subtract the equations.

Solve for one variable.

Then solve for the second.

If the smaller coefficients don't divide evenly into the bigger ones.

For either variable.

It's okay.

That's scenario three.

We'll go over that next.

For now, here are the updated steps.

1) Line up and label the equations (1) and (2)
2) Look at the coefficient of each variable in both equations
    i. If one goes evenly into the other, figure out by what number. Then, multiply the entire equation with the smaller coefficient by that number
3) Label this new equation as (3)
4) Line up the two equations with matching coefficients
5) Add or subtract the equations to eliminate this variable
6) Solve the new equation for the remaining variable
7) Plug this answer back into either original equation to find the other variable

Worked Example 5

Solve the following simultaneous equations using the elimination method.

$$\mathbf{4x + 6y = 10}$$
$$\mathbf{2x - 3y = 11}$$

Explanation	**Solution**	
Write and label the equations.	$4x + 6y = 10$ $2x - 3y = 11$	(1) (2)
Choose the variable x to eliminate.	$\mathbf{2 \times (2)}$	
Multiply each term of the second equation by 2. Label this new equation as (3).	$2 \times (2x - 3y = 11)$ $4x - 6y = 22$	 (3)
Line up the two equations with matching coefficients vertically and label them accordingly.	$4x + 6y = 10$ $4x - 6y = 22$	(1) (3)
Subtract the first equation from the third.	$\mathbf{(3) - (1)}$	
Solve this equation for y.	$-6y - 6y = 22 - 10$ $-12y = 12$ $y = -1$	
	Sub $\mathbf{y = -1}$ **into (1)**	
Plug this value into either original equations and solve for x. (e.g. first equation)	$4x + 6y = 10$ $4x + 6(-1) = 10$ $4x - 6 = 10$ $4x = 16$ $x = 4$	
Write answer.	$\therefore x = 4 \; and \; y = -1$	

Worked Example 6

Solve the following simultaneous equations using the elimination method.

$$\mathbf{5x + 4y = 12}$$
$$\mathbf{3x + 12y = 36}$$

Explanation	Solution
Write and label the equations.	$5x + 4y = 12$ (1) $3x + 12y = 36$ (2)
Choose the variable y to eliminate.	$\mathbf{3 \times (1)}$
Multiply each term of the first equation by 3. Label this new equation as (3).	$3 \times (5x + 4y = 12)$ $15x + 12y = 36$ (3)
Line up the two equations with matching coefficients vertically and label them accordingly.	$3x + 12y = 36$ (2) $15x + 12y = 36$ (3)
Subtract the second equation from the third.	$\mathbf{(3) - (2)}$
Solve this equation for x.	$15x - 3x = 36 - 36$ $12x = 0$ $x = 0$
	Sub $x = 0$ into (1)
Plug this value into either original equations and solve for y. (e.g. first equation)	$5x + 4y = 12$ $2(0) + 4y = 12$ $4y = 12$ $y = 3$
Write answer.	$\therefore x = 0 \; and \; y = 3$

Worked Example 7

Solve the following simultaneous equations using the elimination method.

$$\mathbf{3y - 8x = 7}$$
$$\mathbf{2x - 5y = 11}$$

Explanation	**Solution**
Write and label the equations.	$-8x + 3y = 7 \quad (1)$ $2x - 5y = 11 \quad (2)$
Choose the variable x to eliminate.	$4 \times (2)$
Multiply each term of the second equation by 4. Label this new equation as (3).	$4 \times (2x - 5y = 11)$ $8x - 20y = 44 \quad (3)$
Line up the two equations with matching coefficients vertically and label them accordingly.	$-8x + 3y = 7 \quad (1)$ $8x - 20y = 44 \quad (3)$
Add the third equation to the first equation.	$(1) + (3)$
Solve this equation for y.	$3y + (-20y) = 7 + 44$ $-17y = 51$ $y = -3$
	Sub $y = -3$ into (2)
Plug this value into either original equations and solve for x. (e.g. second equation)	$2x - 5y = 11$ $2x - 5(-3) = 11$ $2x + 15 = 11$ $2x = -4$ $x = -2$
Write answer.	$\therefore x = -2 \text{ and } y = -3$

The Elimination Method 3: When Coefficients Don't Play Nice

Alright.

Scenario three.

The "harder" simultaneous equations.

The coefficients don't match.

And.

The smaller coefficients don't go evenly into the bigger ones.

For these, we can eliminate either variable.

Remember, we need the coefficients to match for one variable.

This time, to make that happen.

We multiply both equations by a number.

Here's what to do.

Pick a variable to eliminate.

Doesn't matter which.

Go for the one with the smaller number.

Multiply each equation by the other equation's coefficient.

The coefficient of the variable you've chosen to eliminate.

Label these two new equations (3) and (4).

From here, you know what to do.

Coefficients now match.

Line up the two new equations.

Add or subtract.

Solve for one variable.

Then solve for the other.

Here's the final updated steps.

1) Line up and label the equations (1) and (2)
2) Look at the coefficient of each variable in both equations
    i. If one goes evenly into the other, figure out by what number. Then, multiply the entire equation with the smaller coefficient by that number
    ii. If they don't divide evenly, for the variable being eliminated, multiply each equation by the other equation's coefficient.
3) Label these new equations as (3) and (4)
4) Line up these two new equations with matching coefficients
5) Add or subtract the equations to eliminate this variable
6) Solve the new equation for the remaining variable
7) Plug this answer back into either original equation to find the other variable

Worked Example 8

Solve the following simultaneous equations using the elimination method.

$$\mathbf{3x + 7y = -8}$$
$$\mathbf{2x + 3y = -2}$$

Explanation	**Solution**
Write and label the equations.	$3x + 7y = -8 \quad (1)$ $2x + 3y = -2 \quad (2)$
Choose the variable x to eliminate.	$\mathbf{2 \times (1)}$ *and* $\mathbf{3 \times (2)}$
Multiply the first equation by 2 and the second by 3.	$2 \times (3x + 7y = -8)$ $3 \times (2x + 3y = -2)$
Label the new equations as (3) and (4) and line them up vertically.	$6x + 14y = -16 \quad (3)$ $6x + 9y = -6 \quad (4)$
Subtract the fourth equation from the third.	$(3) - (4)$
Solve this equation for y.	$14y - 9y = -16 - (-6)$ $5y = -10$ $y = -2$
	Sub $y = -2$ into (2)
Plug this value into either original equations and solve for x. (e.g. second equation)	$2x + 3y = -2$ $2x + 3(-2) = -2$ $2x - 6 = -2$ $2x = 4$ $x = 2$
Write answer.	$\therefore x = 2$ *and* $y = -2$

Worked Example 9

Solve the following simultaneous equations using the elimination method.

$$\mathbf{5a - 2b = 20}$$
$$\mathbf{4a + 5b = 49}$$

Explanation	Solution
Write and label the equations.	$5a - 2b = 20 \quad (1)$ $4a + 5b = 49 \quad (2)$
Choose the variable b to eliminate.	$\mathbf{5 \times (1)}$ *and* $\mathbf{2 \times (2)}$
Multiply the first equation by 5 and the second by 2.	$5 \times (5a - 2b = 20)$ $2 \times (4a + 5b = 49)$
Label the new equations as (3) and (4) and line them up vertically.	$25a - 10b = 100 \quad (3)$ $8a + 10b = 98 \quad (4)$
Add the third equation to the fourth.	$\mathbf{(3) + (4)}$
Solve this equation for a.	$25a + 8a = 100 + 98$ $33a = 198$ $a = 6$
	Sub $a = 6$ into (1)
Plug this value into either original equations and solve for b. (e.g. first equation)	$5a - 2b = 20$ $5(6) - 2b = 20$ $30 - 2b = 20$ $-2b = -10$ $b = 5$
Write answer.	$\therefore a = 6 \text{ and } b = 5$

Worked Example 10

Solve the following simultaneous equations using the elimination method.

$$\mathbf{2y = 9 - 3x}$$
$$\mathbf{-2x - 33 = -3y}$$

Explanation	**Solution**
Line up and label the equations.	$3x + 2y = 9 \quad (1)$ $-2x + 3y = 33 \quad (2)$
Choose the variable y to eliminate.	$3 \times (1)$ *and* $2 \times (2)$
Multiply the first equation by 3 and the second by 2.	$3 \times (3x + 2y = 9)$ $2 \times (-2x + 3y = 33)$
Label the new equations as (3) and (4) and line them up vertically.	$9x + 6y = 27 \quad (3)$ $-4x + 6y = 66 \quad (4)$
Subtract the fourth equation from the third.	$(3) - (4)$
Solve this equation for x.	$9x - (-4x) = 27 - 66$ $13x = -39$ $x = -3$
	Sub $a = 6$ into (1)
Plug this value into either original equations and solve for y. (e.g. first equation)	$3x + 2y = 9$ $3(-3) + 2y = 9$ $-9 + 2y = 9$
	$2y = 18$ $y = 9$
Write answer.	$\therefore x = -3 \text{ and } y = 9$

Your Turn To Try

1 **After eliminating b what's the new equation?**

$$4a - 3b = -10$$
$$2a + 3b = 4$$

A $6a = -6$ B $2a = -6$ C $6a = 14$ D $2a = 14$

2 **After eliminating x what's the new equation?**

$$-5x + 2y = 16$$
$$-5x - 3y = 1$$

A $5y = 17$ B $-y = 15$ C $-5y = 15$ D $5y = 15$

3 **For the following system of equations. What is the value of x?**

$$3x + 4y = 2$$
$$5x + 2y = -6$$

A $x = -2$ B $x = 3$ C $x = 1$ D $x = 2$

4 **What is the value of y for the following the system of equations**

$$-2x + 3y = 7$$
$$3x - 2y = -3$$

A $y = -3$ B $y = 3$ C $y = 1$ D $y = -1$

5 **What is the solution to the system of equations.**

$$3x + 4y = 13$$
$$2x - 5y = 1$$

A $x = 2, y = 3$ B $x = 1, y = 3$ C $x = 3, y = 1$ D $x = -1, y = 3$

Question	1	2	3	4	5
Answer	A	D	A	B	C

Time To Practice

1 Solve the following pairs of simultaneous equations, using the elimination method.

a $3x + 5y = 21$
$2x + 5y = 19$

b $3x + 5y = 17$
$3x - 4y = -19$

c $6a - 4b = 14$
$-2a - 4b = -10$

d $7a + 6b = -8$
$4a - 6b = -14$

e $11p - 4q = 2$
$7p - 4q = -6$

f $3p - 5q = 6$
$-2p - 5q = 46$

g $7x - 3y = -16$
$-7x + 2y = 6$

h $-3x + 4y = 14$
$5x + 4y = -34$

2 Solve the following pairs of simultaneous equations, using the elimination method.

a $2x + 3y = 9$
$4x + 5y = 15$

b $3x + 7y = 1$
$9x - 4y = -22$

c $12a + 3b = 3$
$-4a - 5b = 11$

d $7a - 6b = -9$
$2a + 12b = -30$

e $5p + 2q = -7$
$2p - 20q = 18$

f $2p - 3q = 0$
$5p + 15q = 45$

g $-2x + 5y = -10$
$8x - 3y = 6$

h $-5x - 6y = 4$
$3x - 2y = 20$

3 Solve the following pairs of simultaneous equations, using the elimination method.

a $3x + 2y = 7$
$2x + 5y = 12$

b $-3x + 2y = -7$
$4x - 5y = 7$

c $4a + 3b = -5$
$3a - 2b = 9$

d $-3a + 4b = -2$
$-4a - 5b = 18$

e $7p - 3q = 5$
$-5p + 4q = -11$

f $-2p + 5q = 2$
$9p - 3q = 30$

g $-8x - 3y = 7$
$5x + 2y = -5$

h $-3x - 7y = -9$
$-5x - 4y = 8$

Challenge Yourself

1 Solve the following simultaneous equations using the elimination method.

a $3 + 4x - 3y = 0$
$4 - 2y + 3x = 0$

b $2x = 15 + 5y$
$-7y = 6x - 23$

c $\frac{2m - 5n}{2} = 16$
$2m + 3n = 0$

d $\frac{2p}{3} + 3q = -4$
$5p - 9q = 33$

e $\frac{x + 2}{2} + \frac{y - 1}{3} = 1$
$5x + 2y = -6$

f $\frac{2x}{3} + \frac{y}{2} = \frac{1}{6}$
$\frac{x}{5} + \frac{3y}{2} = -\frac{2}{5}$

5. Answers

The Substitution Method

Time to Practice

1 a $x = 1, y = 1$ b $x = 2, y = 2$ c $x = 2, y = 1$

d $x = -3, y = -2$ e $x = 7, y = 3$ f $x = 12, y = 5$

g $x = 2, y = -5$ h $x = -6, y = -2$

2 a $x = 1, y = 6$ b $x = 9, y = 4$ c $a = 2, b = 5$

d $a = 8, b = 0$ e $p = -3, q = 7$ f $p = 5, q = -10$

g $m = 12, n = 1$ h $m = -2, n = -11$ i $r = 4, s = 8$

j $r = -8, s = -9$

Challenge Yourself

1 a $m = 9, n = 2$ b $x = 13, y = -3$ c $r = 4, s = 0$

d $c = 3, d = 4$ e $p = \frac{4}{3}, q = 1$ f $a = \frac{7}{4}, b = \frac{1}{2}$

The Elimination Method

Time to Practice

1 a $x = 2, y = 3$ b $x = -1, y = 4$ c $a = 3, b = 1$

d $a = -2, b = 1$ e $p = 2, q = 5$ f $p = -8, q = -6$

g $x = 2, y = 10$ h $x = -6, y = -1$

2 a $x = 0, y = 3$ b $x = -2, y = 1$ c $a = 1, b = -3$

d $a = -3, b = -2$ e $p = -1, q = -1$ f $p = 3, q = 2$

g $x = 0, y = -2$ h $x = 4, y = -4$

3 a $x = 1, y = 2$ b $x = 3, y = 1$ c $a = 1, b = -3$

d $a = -2, b = -2$ e $p = -1, q = -4$ f $p = 4, q = 2$

g $x = 1, y = -5$ h $x = -4, y = 3$

Challenge Yourself

1 a $x = -6, y = -7$ b $x = 5, y = -1$ c $m = 6, n = -4$

d $p = 3, q = -2$ e $x = -4, y = 7$ f $x = \frac{1}{2}, y = -\frac{1}{3}$

What did you think of the book?

I really hope this book helped you with Algebra.

The only way I'll know is if you leave a quick review.

So, if you don't mind, could you write one?

Doesn't have to be long.

Positive.

Negative.

What you liked.

What you didn't.

Anything.

That way I'll know if this book actually helped.

and I can work on making it even better.

Thanks heaps.

Made in the USA
Monee, IL
23 October 2025

32696821R00171